Industry 4.0 and Digitization

This book includes studies on regions, industries and tendencies of industrial change and spatial concentration of competences and industrial potentials. The chapters in this volume provide for discussions concerning a wider understanding of situations related to Industry 4.0 and digitization. It also reaches out further than towards technology and economy because it includes regional and metropolitan societies, workforces and the divergencies of effects and opportunities.

Industry 4.0 and digitization are new transformations for regions and metropolises where technologies are applied but regionally can appear as a continuation of innovative processes where it is developed. The divergent presence of competences creates a selectivity process among regions. There are individual industry-location-nexuses formed out of competences of industries, labour force and research which are complemented by public policies providing support towards such adaptation of innovation and change. Regional societies formed from skilled and educated labour become an important basis for participation in innovation and supply chains. Since smart factories widely can be managed remotely, this also shows a concentration of decision making. Simultaneously, it forms a polycentric de-concentration, indicating some more important locations as central within the networks. These systematic changes continue to deepen over time. While public policies may match innovative opportunities at the appropriate moment, they also contribute to a continuation of uneven development and divergent societal tendencies. Industry 4.0 and digitization indicate a wide and selective change of organization associated with new technologies and innovation. While some regions and metropolises can continue to build both innovative competences and innovative societies based on innovative labour force, others will participate because of their position in supply chains.

The chapters in this book were originally published as a special issue of the journal, *European Planning Studies*.

Ulrich Hilpert is Professor of Comparative Government at the University of Jena, Germany; Fellow of the Academy of Social Sciences, London; Senior Fellow of the Hans-Böckler-Foundation, Düsseldorf, and has been visiting professor at a dozen universities in Europe and the United States. His main areas of research are comparative studies in technology, innovation, regional development, global networking and skilled and university trained labour.

Industry 4.0 and Digitization

Regions and Metropolises Facing Divergent Social and Industrial Change

Edited by
Ulrich Hilpert

Routledge
Taylor & Francis Group

LONDON AND NEW YORK

First published 2022
by Routledge
4 Park Square, Milton Park, Abingdon, Oxon OX14 4RN

and by Routledge
605 Third Avenue, New York, NY 10158

Routledge is an imprint of the Taylor & Francis Group, an informa business

British Library Cataloguing in Publication Data
A catalogue record for this book is available from the British Library

ISBN: 978-1-032-27304-4 (hbk)
ISBN: 978-1-032-27305-1 (pbk)
ISBN: 978-1-003-29219-7 (ebk)
DOI: 10.4324/9781003292197

Typeset in Minion Pro
by Newgen Publishing UK

Publisher's Note
The publisher accepts responsibility for any inconsistencies that may have arisen during the
conversion of this book from journal articles to book chapters, namely the inclusion of journal
terminology.

Disclaimer
Every effort has been made to contact copyright holders for their permission to reprint material
in this book. The publishers would be grateful to hear from any copyright holder who is not here
acknowledged and will undertake to rectify any errors or omissions in future editions of this book.

Contents

Citation Information   vii
Notes on Contributors   ix

Introduction – spatial evolution in the light of innovative transformation:
the impact of policies and institutions in divergent situations   1
Ulrich Hilpert

1   Regional selectivity of innovative progress: Industry 4.0 and digitization ahead   9
    Ulrich Hilpert

2   The impact of Industry 4.0 on supply chains and regions: innovation in the
    aerospace and automotive industries   26
    Desmond Hickie and James Hickie

3   The transition of regional innovation systems to Industry 4.0: the case of
    Basque Country and Catalonia   42
    Francesco D. Sandulli, Elena M. Gimenez-Fernandez and
    Maria Isabel Rodriguez Ferradas

4   The growing inequalities in Italy – North/South – and the increasing
    dependency of the successful North upon German and French industries   57
    Matteo Gaddi, Nadia Garbellini and Francesco Garibaldo

5   How does Industry 4.0 affect the relationship between centre and periphery?
    The case of manufacturing industry in Germany   76
    Samuel Greef and Wolfgang Schroeder

6   Glowing cities and the future of manufacturing in the US and Europe:
    how digitalization will impact metropolitan areas depending on sectoral
    dominances and regional skill distribution   92
    Yasmin M. Hilpert

7   The Korean approach to Industry 4.0: the 4th Industrial Revolution from
    regional perspectives   110
    Sunyang Chung and Jiyoon Chung

8   Industry 4.0/Digitalization and networks of innovation in the North
    American regional context   128
    Paul M.A. Baker, Helaina Gaspard and Jerry A. Zhu

9   Industry 4.0 as a 'sudden change': the relevance of long waves of economic
    development for the regional level   143
    Walter Scherrer

    Conclusion – the complexity of reorganizing industries and value chains:
    challenges and selectivity of Industry 4.0 and digitization   158
    Ulrich Hilpert

    Index   163

Citation Information

The following chapters were originally published in the journal *European Planning Studies*, volume 29, issue 9 (2021). When citing this material, please use the original page numbering for each article, as follows:

Introduction

Spatial evolution in the light of innovative transformation: the impact of policies and institutions in divergent situations
Ulrich Hilpert
European Planning Studies, volume 29, issue 9 (2021), pp. 1581–1588

Chapter 1

Regional selectivity of innovative progress: Industry 4.0 and digitization ahead
Ulrich Hilpert
European Planning Studies, volume 29, issue 9 (2021), pp. 1589–1605

Chapter 2

The impact of Industry 4.0 on supply chains and regions: innovation in the aerospace and automotive industries
Desmond Hickie and James Hickie
European Planning Studies, volume 29, issue 9 (2021), pp. 1606–1621

Chapter 3

The transition of regional innovation systems to Industry 4.0: the case of Basque Country and Catalonia
Francesco D. Sandulli, Elena M. Gimenez-Fernandez and Maria Isabel Rodriguez Ferradas
European Planning Studies, volume 29, issue 9 (2021), pp. 1622–1636

Chapter 4

The growing inequalities in Italy – North/South – and the increasing dependency of the successful North upon German and French industries
Matteo Gaddi, Nadia Garbellini and Francesco Garibaldo
European Planning Studies, volume 29, issue 9 (2021), pp. 1637–1655

Chapter 5

How does Industry 4.0 affect the relationship between centre and periphery? The case of manufacturing industry in Germany
Samuel Greef and Wolfgang Schroeder
European Planning Studies, volume 29, issue 9 (2021), pp. 1656–1671

Chapter 6

Glowing cities and the future of manufacturing in the US and Europe: How digitalization will impact metropolitan areas depending on sectoral dominances and regional skill distribution
Yasmin M. Hilpert
European Planning Studies, volume 29, issue 9 (2021), pp. 1672–1689

Chapter 7

The Korean approach to Industry 4.0: the 4th Industrial Revolution from regional perspectives
Sunyang Chung and Jiyoon Chung
European Planning Studies, volume 29, issue 9 (2021), pp. 1690–1707

Chapter 8

Industry 4.0/Digitalization and networks of innovation in the North American regional context
Paul M.A. Baker, Helaina Gaspard and Jerry A. Zhu
European Planning Studies, volume 29, issue 9 (2021), pp. 1708–1722

Chapter 9

Industry 4.0 as a 'sudden change': the relevance of long waves of economic development for the regional level
Walter Scherrer
European Planning Studies, volume 29, issue 9 (2021), pp. 1723–1737

For any permission-related enquiries please visit:
www.tandfonline.com/page/help/permissions

Notes on Contributors

Paul M.A. Baker, Center for Advanced Communications Policy (CACP), Georgia Institute of Technology, USA.

Jiyoon Chung, Department of Global Commerce, Soongsil University, Seoul, Korea.

Sunyang Chung, Department of Technology Management, Konkuk University, Seoul, Korea.

Matteo Gaddi, Fondazione Claudio Sabattini, Bologna, Italy.

Nadia Garbellini, Department of Linguistic and Cultural Studies, Università degli Studi di Modena e Reggio Emilia, Modena, Italia.

Francesco Garibaldo, Fondazione Claudio Sabattini, Bologna, Italy.

Helaina Gaspard, The Institute of Fiscal Studies and Democracy (IFSD), Ottawa, Canada.

Elena M. Gimenez-Fernandez, University Complutense of Madrid, Madrid, Spain.

Samuel Greef, Political System of Germany, University of Kassel, Kassel, Germany.

Desmond Hickie, University of Chester, Chester, UK.

James Hickie, Queen's Management School, Queen's University, Belfast, UK.

Ulrich Hilpert, Department of Social Sciences, Friedrich-Schiller University, Jena, Germany; Hans-Böckler-Foundation, Düsseldorf, Germany.

Yasmin M. Hilpert, Senior Director of Policy, Council on Competitiveness, Washington, DC, USA.

Maria Isabel Rodriguez Ferradas, Tecnum – University of Navarra, Pamplona.

Francesco D. Sandulli, University Complutense of Madrid, Madrid, Spain.

Walter Scherrer, Department of Economics, University of Salzburg, Salzburg, Austria.

Wolfgang Schroeder, WZB Berlin Social Science Center, Germany.

Jerry A. Zhu, The Institute of Fiscal Studies and Democracy (IFSD), Ottawa, Canada.

Introduction – spatial evolution in the light of innovative transformation: the impact of policies and institutions in divergent situations

Ulrich Hilpert

The discussion and perception of digitization and Industry 4.0 are widely concentrated on its technological impact and applications. Algorithms, artificial intelligence, the robots, remote steering of manufacturing plants, 3D printing, computing, new services, etc. are the dominant focus and point to a new future. In addition, there are discussions about the impact on employment, work, changes to industries and services, and what kinds of skills may become obsolete, or what new skills and competences will be in demand. New occupations will emerge while others may disappear. The overall picture and expectations are varied, and the various perceptions of the future are controversial. Similarly, the effects are highly divergent when regional situations are considered. There are regions which are the home of software developers, and others are locations where new technologies are applied. In addition, industrial structures, skills and education, supply chain structures, the products manufactured, the services provided and the composition of regional economies, refer to a wide range of divergent impacts of new Industry 4.0 technologies. Existing situations, their context and relationships, characterized through value chains, are often decisive and have a strong impact on the reorganization and management of the enterprises and supply chains which are critical to individual regions and metropolises.

This special issue contains contributions which shall help to build a better understanding of such new processes and why they are, of necessity, empirically highly divergent. The contributions cover a number of different issues and indicate that they may demonstrate divergent phenomena, although they refer to a common underlying rationality. The technologies related to industry 4.0 and digitization are as divergent as the regional situations they meet. But some regions may benefit more from these changes than others and some may even have problems participating at all. Since a modern, high-speed internet is a fundamental condition to participate in these changes and transformations, clearly many regions are left behind while metropolises are privileged. Agglomerations have better access to international communication and transportation, and providing the infrastructure necessary for modern industries and services is much more frequent in agglomerations than in peripheral regions. Nevertheless, the existing situations in industries, services, research and the labour force complemented by appropriate public policies contain the potential for uneven development. These can appear within countries across regions and may vary between metropolises and continents. Consequently, opportunities vary a lot but they refer to the same dynamism and rationality. Thus, empirical phenomena may refer to different industries, capabilities and situations and they may also involve technology developers or appliers.

Thus, it is important to integrate a rich diversity of examples and a comparative view, which both help to understand such processes as a system and over a longer period. The contributions contain examples from Spain (Catalonia and Basque Country), Italy (comparing the North East and North West, Centre and the South), Germany (Baden-Württemberg, Brandenburg, North Rhine-Westphalia), the US (Massachusetts) and Canada (Ontario) and Korea (Greater Seoul). In addition, locations are considered either as the 100 largest metropolises in the US compared with the European situation or when industries are considered which integrate locations into supply chains or value chains (aircraft, automobile etc.). Occasionally, reference is given to California, the United Kingdom, Detroit, or Washington State to illustrate such processes and examples of supply chains. This also includes the importance of governmental systems (see Germany, Spain, Italy, Korea, the US or Canada) and intermediary institutions and the relationship with services (see Massachusetts and Ontario). Services may change independently and they may also change when related with manufacturing or process industries (see Catalonia, the Basque Country, Italian and German regions, the different types of metros in the US and Europe and how these emerge in relation to aerospace, or over a longer period).

The changing situations of supply chains is also indicated when the remote steering of plants by smart factories is considered (see Italy and Korea) and how this impacts on management. While activities had to be organized and determined in individual plants, Industry 4.0 and digitization allows the planning of processes and variations during manufacturing, as well as control of machines, robots and other advanced equipment. In addition, inventories can be managed remotely, and robots can supply material from stock. Similarly, accounting and arranging for changes in processes can also be managed remotely. Consequently, many activities which were controlled by local management disappear when more can be organized virtually. Opportunities for data management help to move further in this direction and can even cope rather flexibly when special customer demands need to be satisfied. Geographical distance or different time zones decreasingly matter when processes and manufacturing can be arranged from far away. Also, when specific human competences are required because of specific situations and tasks required by particular products, this is not necessarily something that needs to be decided by management in the plant. This changes the working conditions, and the economic rationality of having the machines running all day long for the entire week means that employees need to be available according to the schedule whereby these new machines need to be programmed, re-arranged, or serviced. Such changes of situations and working conditions, of course, are closely associated with the locations where new equipment of Industry 4.0 is applied. More individual competences and human capabilities are required in regions where Industry 4.0 equipment is designed, developed, and manufactured and thus changes of working conditions and everyday life are less frequent or intense.

The increasing application of Industry 4.0 equipment and digitization of processes have a deep and far-ranging influence. While these issues are widely discussed regarding changes in manufacturing and processes or concerning skills, education and employment, the effects on the reorganization of supply chains and value chains have important impacts on regions and metropolises. The potential of industries, research and labour form divergent arrangements at locations which respond differently to such changes. Since both Industry 4.0 and digitization are increasingly important as part of continental and global value chains this is already rather selective. In addition, there are differences between enterprises and locations where equipment and software are developed and those where it is applied. Thus, the kind of participation in Industry 4.0 and digitization is widely determined by the position along the value chains. There are opportunities for suppliers and their locations to continue their development and modernization according to their relationship to the final OEMs. While they participate in the processes of innovation by advanced manufacturing, simultaneously, they now depend systematically on the final OEMs who use their products and services.

Consequently, there is uneven development even among regions which develop or employ such new equipment. Even progressive innovative development is divergent and is related to regions and metropolises according to their divergent situations. It is important to understand that the supply chains and value chains are highly heterogeneous and dependent positions within such networks refer to divergent competences. Specific competences can be tailored much more and so contributions to the value chains are highly divergent. It is the ways in which such divergent competences match the needs of value chains that are decisive for the level of development related to particular opportunities and processes of innovation. The chains are undergoing reorganization and the ways that this is experienced in regions and metropolises is based on their individual situations. Thus, skills, education, traditions and orientations in manufacturing or processing industries, as well as in services, create a situation and context which is fundamental for the kind of contribution to be made and how this is organized. These far-reaching conditions, and how they are arranged regionally and in metropolises, make Industry 4.0 and digitization a process which is indeed new, and has a deeper impact than just the introduction of new equipment. This is what this special issue wishes to express in the contributions focusing on specific elements and situations of the entire process in a comparative way.

The far-ranging aspects of the individual processes of development and how this relates with divergent regional development is discussed in the contribution by Ulrich Hilpert. It is important to see the rich diversity of relationships within Industry 4.0 and digitization. While it is discussed frequently, how a region may develop and what the effects of new technological opportunities are upon it, here it is the aim to develop the knowledge about enterprises, research capabilities, value chains and skill requirements into an understanding of where and how this may generate effects on socio-economic development. The innovative processes are realized alongside the value chains and, according to their regional or metropolitan advantages and competences, then locations can participate. Thus, this contribution argues that there needs to be a match between both industrial opportunities and innovative requirements. This matching is also required between these newly developed technologies if they are to be applied in other regions which are also faced for the first time with these changes in the technologies, organization and skills and education required of them. Thus, it is pointed out that there are different timelines of application to be established alongside the reorganization of value chains. Capabilities will already be established that will continue for a longer period and which will be resistant to fundamental changes. New ways of running plants within the supply chain by remote management and the transfer of information for 3D printers etc. allow a virtualization of space and replace proximity by the use of appropriate equipment at geographically dispersed locations. While such changes may appear rather sudden in some regions where enterprises introduce them, in fact, they represent continuities in the contexts of the design and development of technologies and of management strategies. Consequently, the paper argues that there is a clear tendency for uneven development based on structures which are associated with Industry 4.0 and digitization.

The important role of industries organized within supply chains and value chains is highlighted by the contribution of James Hickie and Desmond Hickie. They choose the cases of two highly globalized manufacturing industries indicating how these are organized and what structures currently exist. Understanding the rationality of both industries and how they may correspond to one another in applying new technologies, it becomes clear why certain regions are contributing to value chains and will continue to do so. The adoption of new technologies in the areas of Industry 4.0 and digitization cannot be understood without a clear picture of the importance of secure supply and how such new technologies may correspond with the situation of the enterprises at a particular place. This includes industrial competences, skills and education, and the potential to implement these changes into their manufacturing base. When comparing aircraft and automobile industries regarding their chains they highlight the similarities and the divergencies which relate to the complexity of the products. Such conditions become important for regions to continue their position within value chains, or for new regions to engage based on windows of opportunities which have opened and which match their capabilities and enterprises.

Arranging for appropriate situations becomes an increasingly important area of public policy. In governmental systems with strong regional governments, there are particular initiatives and attempts by regional administrations. The contribution by Francesco Sandulli, Elena Gimenez-Fernandez and Maria Rodriguez Ferradas address Catalonia and the Basque Country as important industrial centres for which they discuss activities in the light of the existing structures. They show that there is a need for appropriate programmes which match their situations and which allow for attractive positioning within supply chains. The cases selected point to the role of institutional structures which suit the local enterprises and their opportunities to make the regional industries fit for Industry 4.0 and digitization. It becomes very clear that having capable institutions alone will not provide for the innovative processes required. They refer to the exchange of knowledge among the agents with capabilities in these new subjects. When they highlight the process as driven strongly by engineering, they also indicate that there are opportunities for regions which may not have leading edge research but can build upon a strong tradition in technology application and engineering. While doing so, they also highlight that the individual regional policies can be very diverse while effectively exploiting these new opportunities.

The integration into European supply chains and value chains is an important context when new equipment based on Industry 4.0 is employed and new opportunities related with digitization are applied is shown by Matteo Gaddi, Nadia Garbellini and Francesco

Garibaldo with regard to the highly diverse situation in Italy. There are the situations of North and South in Italy, and the problems of vitalizing the country's disadvantaged region in the South which are challenged by processes associated with these innovations. Although there is the highly regarded technical university of Bari, there is little innovative activity which may change the situation of the surrounding regions. Thus, it may be difficult to establish competitive supply chains. In contrast, the authors show the uneven effect of national policies because they do meet the needs and circumstances of enterprises and regions in Northern Italy. The success which may be achieved, in the end, again may help to continue the country's uneven development. On the other hand, the success of the Northern regions is widely dependent on supplying French and German industries. In doing so, and facing a trend towards smart factories which may be run by remote management, the paper points to the lack of attention the national policies pay to both skilled labour and to the broader societal effects even in such more modernized regions.

The importance of the full arrangement is expressed by Samuel Greef and Wolfgang Schroeder when they highlight both the existing industrial competences and the capabilities of regional governments. When they analyse the individual German Länder it turns out very clear that there are strong opportunities where the industrial structures are modern and open to innovations related with Industry 4.0 and digitization. They show very clearly how Bavaria, Baden-Württemberg and Northrhine-Westphalia match the new technologies and how they are ready to continue their position. In addition, strong policy programmes generate support to develop new technologies, apply new opportunities with existing products or to integrate new equipment into manufacturing processes. In contrast, there are rather limited opportunities to benefit from such changes when the Länder are industrially rather weak and public budgets are limited. National programmes and matching Länder programmes, in general, suit the industries ready for these innovations but clearly there is a tendency towards geographic concentrations. The authors can show how these processes and policies are contributing to uneven regional development characterized by innovative opportunities.

It is important to see that such uneven regional effects of Industry 4.0 and digitization are also of significant impact on metropolises with different structures. Since the effects are different because the situations are divergent, Yasmin Hilpert uses a comparative approach concerning metropolises in Europe and the US. The typology of metropolises she develops allows for a precise analysis of the data to compare the situations. Consequently, there are divergent demands for knowledge and differences concerning jobs which may fall away or will be created. Similarly, the author shows that requirements for skills and education create divergent agglomerations of talent and indicate the processes of development which are different between the types of metropolises. Finally, the European-US comparison shows that the effects of Industry 4.0 and digitization are different because of both the mix of industries and the higher average skill level in Europe. This indicates that experiences from different places can be used to develop a better understanding, but the context needs also to be considered. Thus, it is shown that Europe will have a different situation to the US and the typology will show divergencies among individual metropolises. Traditions in advanced manufacturing and a highly skilled labour force will help to avoid processes of societal polarization which is frequent in metropolises characterized by old industries and which lack a mix of industries.

The strong concentration of innovation related with Industry 4.0 and digitization follows from the advantage which metropolises provide when it comes to research, skills, education and infrastructure. While these areas of innovation clearly require appropriate situations to flourish, Sunyang Chung and Jiyoon Chung point to the effects across regions when policies are not designed to allow for participation from less central regions. The Korean strategy particularly aims at a diffusion of smart factories, but there is a concentration predominantly on Seoul, Gyeonggi and a few other regions of the country. These are the places which match best with the requirements to realize industry 4.0 and digitization but they also point to uneven regional opportunities as has already been identified in other countries. Metropolitan areas draw these firms and competences because they are appropriate locations for firms to modernize through applying these opportunities. Again, a relationship with the governmental structures and policy design becomes clear, when the authors note that a difficulty to successfully implement Industry 4.0 relates to the fact that regional governments have not been included in the 'Presidential Committee on the Fourth Industrial Revolution' (PCFIR). This again indicates how many different types of actors and activities need to be coordinated when launching this innovation.

The importance of the regional level is shown by Paul Baker and Helaina Gaspard when contributing their comparative study on software sectors in Massachusetts, USA, and Ontario, Canada. In addition to other studies on regions provided in this special issue, they pay particular attention to regional innovation policy which enables developments that help to build knowledge clusters. They highlight the role of interaction between the actors engaged in the region and, thus, demand for the contribution of intermediaries within regional networks. It is shown that structures are important, but activities are required to get processes running. The software example also indicates both that modern industries are closely linked with high-level services, and it shows how far-reaching Industry 4.0 and digitization are. Again, existing capacities of the regions are important and they play an enabling role when the stakeholders engage. As already noticed from the cross-sectoral impact of the new technologies, collaborations of actors are important to raise the potential of the regions based on intra- and intersectoral networks. When the authors show how important networks of knowledge and intermediaries are, it also becomes obvious that regions are in an advantaged situation to benefit from new opportunities when they already have established institutional contexts.

As could be learned through all these contributions, existing situations are important for future opportunities – to participate in innovation which is based on Industry 4.0 and digitization. Since the situations are varied there are also different perceptions of technologies. Walter Scherrer points out that whether technologies are developed in a region or applied makes a fundamental difference to the way they are perceived, which may lead to a misunderstanding of the techno-economic paradigm. While the developers of the technologies and new robots and equipment regard this as an advanced but consequential continuation of their research and design, those who buy and apply these technologies focus on the new opportunities and the change in organization and management they are facing. Clearly, the latter perspective and experience is one of a sudden change and the prevailing techno-economic paradigm is not perceived clearly. The author makes an important point by applying the long waves approach, which is more than just pointing to longer time periods, but it shows how important a full picture is to understand processes of innovation. What is demonstrated clearly is that, alongside the long waves, there are more detailed and systemic studies needed. The wide-spread applications from manufacturing and process industries to services and management, of course, make the analysis more complex.

As the contributions show, when analysing Industry 4.0 and digitization it is important to understand the situation and the context. While the existing capabilities and structures in industry and research are already well discussed these technologies, based on previous research and related to a wide range of opportunities, indicate that regional effects as well as innovation processes need to be understood in the context of processes which take place elsewhere. This helps the understanding of innovation more completely and it shows the position which individual enterprises and regions can fill. Value chains take advantage of divergent competences and integrate those as the final product put together. Global and increasingly European value chains are characterized by the divergent innovative contributions and improving services and software development at different places which contribute to them by making improved products available through the internet infrastructure. Thus, processes related with Industry 4.0 and digitization demand for more comprehensive analysis which includes these different topics.

The regional arrangement of research, industry and services, of course, varies according to the contribution they make within supply chains. The regions and metropolises where most advanced processes of innovation are located need a fully developed context which is complemented by access through the internet to capabilities from elsewhere and is home of the skilled labour force required. Increasing competition globally, or more specifically from south of the Mediterranean Sea, demands change and adjustment from regions which supply less complex products into supply chains. While these need to advance their capabilities to retain their competitiveness, there is also the opportunity to re-Europeanize supply chains since the new technologies of Industry 4.0 can also reduce the share of labour costs and make European locations attractive once again. It is not yet clear, in which areas such re-shoring will be realized, and whether, on the other hand, a combination of new technologies with cheap and unskilled labour might allow for new tendencies towards off-shoring.

Clearly, there will be activities to modify or change supply chains and value chains based on new technological opportunities. The different tendencies will impact on the situations of regions and metropolises in Europe and globally. As the contributions have shown, the effects also depend greatly on the skills and education of the work force. In many European regions initiatives to improve the levels of skill and the existing mix of industrial structures will help to balance the effects of these new technologies by creating a rich diversity of research capabilities and utilizing a wide range of traditional competences. The contributions to this special issue show that regional or metropolitan situations which can relate to a capable labour force from highly skilled blue-collar labour to well-educated knowledge workers can find themselves in a beneficial situation when it comes to innovation processes and they are also less affected by the risks of remote management related with smart factories. They are ready to create complex products and processes and be at the forefront of applications, or even engaging development and design. Consequently, they find attractive positions within the long waves and the longer the waves last, the more sustainable these arrangements might become. Public policies which include regional governments and allow for contributions from all stakeholders and engage institutions which address competent labour forces, as well as support of enterprises, are important for the realization of innovation based on Industry 4.0 and digitization. There is a need for much more research to understand these matters.

Last but not least we need to thank the 'Hans-Böckler-Foundation, Düsseldorf, Germany', for making a face-to-face workshop possible, which has helped our exchange of ideas and views and created a synergy throughout the group.

Regional selectivity of innovative progress: Industry 4.0 and digitization ahead

Ulrich Hilpert

ABSTRACT

Industry 4.0 and digitization are new transformations for regions and metropolises where technologies are applied but regionally can appear as a continuation of innovative processes where it is developed. The divergent presence of competences creates a selectivity process among regions. There are individual industry-location-nexuses formed out of competences of industries, labour force and research which are complemented by public policies providing support towards such adaptation of innovation and change. Regional societies formed from skilled and educated labour become an important basis for participation in innovation and supply chains. Since smart factories widely can be managed remotely, this also shows a concentration of decision making. Simultaneously, it forms a polycentric de-concentration, indicating some more important locations as central within the networks. These systematic changes continue to deepen over time. While public policies may match innovative opportunities at the appropriate moment, they also contribute to a continuation of uneven development and divergent societal tendencies. Industry 4.0 and digitization indicate a wide and selective change of organization associated with new technologies and innovation. While some regions and metropolises can continue to build both innovative competences and innovative societies based on innovative labour force, others will participate because of their position in supply chains.

New technologies raise different issues in different regional contexts. In some contexts, a technological advance may create great opportunities. In another, it may pose severe challenges. In a third, it may simply be inapplicable, and so be irrelevant. Sometimes, a region may adapt its existing economic and social situation to adopt a new technology, taking its opportunities and ameliorating its challenges. It may even need to do so using a variety of different approaches and strategies. Other regions may have the advantage of developing the new technologies (rather than just accommodating them), or maybe able to apply new technologies to develop new products. Finally, a strong research base may attract wholly new economic or industrial capabilities to a region, metropolis or location, which did not have them before. Thus, new technologies can be related to industrial restructuring, highly innovative processes or start-ups which engage in R&D but do not depend directly on manufacturing to add value and achieve economic success. In this respect, the discussion on Industry 4.0 and digitization is very different from previous processes of innovation and the emergence of new technologies. While biotechnology or new materials clearly address the needs of particular industries and thus are influential on particular regions, with regard to Industry 4.0 and digitization, there can be an impact on almost any sector of manufacturing, many services used in everyday life or business, the expansion and reorganization of value chains and the manipulation of large data or algorithms.

Hence, experiences are as divergent as the wide range of regional and sectoral contexts within which these new technologies can be applied and implemented. Transportation, manufacturing or data processing vary according to industrial sectors, urban situations and value chains in which digital technologies are applied. Consequently, contributions by Industry 4.0 and digitization vary widely and can help in very different processes of development – in different places and with different effects. Thus, regions are significant as a basis for strong development and as expressions of the divergencies in such innovative processes. While considering these divergencies, competences in research are important and frequently supported by public policies. Governmental structures, which are characterized by centralized or federal structures, are significant and the concentration of especially high expertise at certain metropolises or regions is important for participation in network-driven innovation.

Thus, research on Industry 4.0 and digitization provides deeper knowledge on the regional effects of the socio-economic processes associated with the technologies and their applications. In addition, it allows for a deeper understanding of divergent systematics of innovation and about the interrelatedness of such processes. This also helps us to understand how much the application of new technological opportunities also relate to government programmes which support research and engineering and create a labour force with skills and education and so allow regions to benefit more fully from such newly emerging opportunities. The divergent situations formed by economic, industrial and socio-political variables indicate the opportunities and limitations of organizing such development. Furthermore, supply chains and the importance of value added from different sources in research, existing industrial competences and labour skills which meet the requirements of the new technologies, also indicate the relationship which exists, or can be built, with regard to economic and technological development of certain regions or metropolises. Profiles of such locations may change over time. They may be included into continental and global networks. New suppliers of knowledge, of goods and services from locations which have no history in such technologies may emerge. Consequently, Industry 4.0 and digitization indicate the regionalization of such processes which necessitate the reorganization of production chains and network-based innovation, as well as its management. In addition, it creates both the unexpected changes at particular regions and the opportunities of innovation in more traditional industries.

Industry 4.0 and digitization: regionally divergent adaptation to new opportunities

Although technological developments as such are not related with regions, nevertheless, regions demonstrate particular competences and relationships with newly emerging technologies (Bellini and Pasquinelli 2018). In addition, of course, there is a clear difference between developing new technological ideas and manufacturing new technological equipment on the one hand, and the application of such opportunities on the other. Consequently, some regions are the home of the competences needed to develop these new technologies, such as big data manipulation, algorithms and augmented reality, while others are associated with their application in manufactured products such as robots, equipment for transportation and logistics. Frequently, opportunities arising from digitization provide a nexus between technical equipment and software. Mobile devices and local detection technologies provide a basis for smart factories, new opportunities for medical services and for technical assistance, despite physical distance.

This also has a strong impact on the organization of manufacturing and the exploitation of divergent opportunities. Since original equipment manufacturers (OEMs) are widely based on supply chains, the management of different contributions and contributors has become highly critical. This is, even more, the case when it concerns highly complex elements and the incorporation of very high-quality processes or components into a final product. Thus, there is a convergent need for technologically advanced products, continuing quality, rapid transportation, on time delivery and reliable supply chains (Hickie 2020). In addition, there is the need to achieve this at low costs. While facing these new challenges, new technologies are applied in manufacturing, services and management. Smart factories, as an example, can be steered and managed remotely to a far greater extent than hitherto (see the contributions by Chung, Scherrer and Gaddi et al.). Timing, responses to changes in market demand, logistics and decisions about product lines, or when to modify products, can be subject of decisions taken far away and managed remotely.

Technological progress towards augmented reality assists the optimization of processes and services, as do new opportunities for visualization, which allow the handling of problems without being physically present. Local detection technologies can manage transportation facilities for the optimal use of the equipment and to reduce transportation times. In combination with 3D printer technologies and new materials, calculations of costs and availability have changed. The transfer of information and knowledge for such 3D printers provides a basis for services at locations where skilled and educated personnel is available at lower costs, where costs of transferring actual parts or personnel to other places are higher (see the contribution by Hickie and Hickie). Also, measurements for suits or shoes can be taken and transferred to a place where labour costs are low, while the transportation of the final product needs to be fast and easy. In addition, the simplification of some products will allow the contribution of labour costs to be reduced to the extent that cheap labour and weak regulation may become less important when compared with the security of supply and closeness to customers.

Consequently, data manipulation, the interconnectedness of machines and robots and optimal logistics in relation with the Internet of Things (IoT) will change structures, processes and contexts, and will be driven by algorithms for optimizing such processes. Such challenges to existing processes and value chains are the consequence of the need to reorganize in the light of Industry 4.0 and digitization. Regions, locations and metropolises participate in such processes and are embedded in their associated networks according to their capabilities, advantages, strengths and traditions. Since the new world of these technologies is not open to any location, it will be highly selective concerning integration into such advanced processes (see the contributions by Sandulli and Gimenez Fernandez, Hickie and Hickie and Baker et al.). Thus, understanding the underlying dynamics of these processes includes the complex relationship between manufacturing, R&D, services and how to manage these in the light of value chains, costs, reliability of supply and quality whilst manufacturing marketable products and affordable prices.

While decisions are taken on activities at locations making specific contributions to supply chains and value added, it is necessary to cope with both the changing division of labour within these processes and its multi-dimensional conditions [What are these. It is not clear?]. This is essential when it comes to being included into, or excluded from, these newly reorganizing transregional networks. Again, opportunities are divergent for regions or metropolises where the new technologies are developed, manufactured and put in place because of their ability to undertake complex processes of management and organization compared to those which are included because new technologies can be introduced and applied because they enjoy favourable transport links with suppliers and/or potential markets. Thus, Industry 4.0 and digitization clearly suggest that this is more than just about manufacturing and new technologies. It clearly includes innovations in management and organization to allow for appropriate supply chains and the remote steering of smart factories and services (see the contributions by Hickie and Hickie, Scherrer and Gaddi et al.). While doing so, this may include new locations as well as those which already contribute to supply chains, which may become reorganized with less autonomy and weaker local management control.

These tendencies indicate highly divergent processes of innovation (Lawton Smith and Assimakopoulos 2020). Some regions or metropolises continue with their outstanding competence in R&D by engaging in further strengthening these established competences. Others are facing the introduction of new technologies which may help them to retain their positions within reorganized value chains, but which may induce increased dependency on decisions taken elsewhere. Meanwhile, new technological opportunities may also create new opportunities for regions without a particularly competent manufacturing labour force. Consequently, the perception of such technologies is divergent from previous situations and capabilities. Where there is a pure application, of course, such technologies usually emerge newly [Not clear]. The wide-ranging consequences associated with Industry 4.0 and digitization give the impression of a rapid change which introduces fundamentally new situations perceived as disruptive changes. While in regions where these technologies are newly applied, they create the impression of disruption, by contrast in regions or metropolises where these new technologies are initially developed, one finds that they consecutively exploit strong competences in R&D which clearly deliver a wide range of advances and opportunities making a strong impact on manufacturing, management and organization globally.

Regionalization by competences: the industry-location-nexus

The technologies of Industry 4.0 and digitization are developed on the basis of earlier competences and can be applied wherever they are needed. The regional effect of these innovations, consequently, closely relates to established industrial situations whose man-ufacturing and R&D competences provide a basis for the development and application of digital technologies. The application of such technologies, in particular, is achieved in manufacturing industries which expect benefits from innovative equipment and services with regard to quality, costs or advanced products. Hence, the regional impact of these new technologies widely follows from existing industrial structures and situations which can take advantage of such opportunities (see the contribution by Scherrer). Con-sequently, it is a continuation and intensification of processes of regionalization which now relate to Industry 4.0 and digitization. When existing clusters are modernized and smart factories are established at these locations, there is a need for good transpor-tation to distribute their products. Digitization means that proximity to transportation facilities becomes even more important. The transmission of information for specifically customized consumer products, or for certain numbers of particular industrial products to be applied or installed (e.g. robots), is related to their immediate availability, which necessitates appropriate infrastructures. The technological opportunities are selectively available among the locations in question, and their transportation needs indicate gen-erally there will be a closeness to metropolitan areas.

While new technologies are always associated with change and with new contributors to R&D, as well as with new suppliers to value chains, the need for transportation and infrastructure continues and suggests both agglomerations of competences and indus-tries and to suitably competent metropolises (see the contribution by Hilpert; Kerr et al. 2017). Capabilities in R&D can be built, but it takes time to become internationally recognized and to participate in the exchanges of innovative personnel which make them attractive partners in research networks (Hilpert and Lawton Smith 2014). This can easily take a decade and is dependent on the research facilities provided and research budgets spent. New suppliers can enter such networks faster by applying new technologies (as in Industry 4.0 and digitization), but they can do so in an economically attractive way only if they are able to manufacture more complex products. Again, this requires skilled and educated labour (Chung 2019; Vassiliadis and Hilpert 2020) which can easily take seven years to a decade to come on stream in the numbers needed to participate effec-tively. No matter whether such processes are based on high-performance manufacturing or on high-tech research, this indicates that the development and production of advanced products takes time. It takes time before emerging innovative locations will be regarded as frontrunners, and thus gain strong, competitive positions in markets.

While Industry 4.0 and digitization allows for many new opportunities, these technol-ogies predominantly address existing industries, markets and products, which are already associated with particular firms and their competences to supply markets or to fit into value chains. Given enterprises are continuously orientated towards constant improve-ments in quality and the innovative capability of their products. In addition, they pay particular attention to the reliability of supply and the existing integration of innovation chains and to their longer term strategies (e.g. Aircraft Industries, Automobile, metal-working industries). Again, organization, culture and management capabilities matter even more in the light of such fundamentally important new technologies. The emer-gence of new locations is frequently associated with new technological opportunities, but rarely will it bring about the disappearance of established locations. In contrast, the new and additional competences and opportunities for manufacturing will increase the number of contributors to innovative processes and, thus, supply chains are becoming more diverse with regard to their numbers of participants. The complex organization of supply, chains including third and fourth-tier suppliers, are highly dependent on quality products and on-time delivery (see the contributions by Hickie and Hickie, Gaddi et al., Baker et al. and Sandulli and Giminez-Fernandez). With the OEMs, there will be an accumulation of effects from innovations contributed throughout the supply chain, which include innovations made in design and manufacturing. This relationship between OEMs and suppliers is reinforced and strengthened by the application or introduction of new innovative opportunities and can leave few opportunities simply to replace successful existing suppliers within a supply chain. Where opportunities do arise for new entrants into the supply chain is to provide new parts or services, or sometimes to supply parts of low complexity.

This relates both to the competence of regions or metropolises and the complexity of the products. The ability to apply the opportunities of Industry 4.0 and digitization helps locations to continue as a base of valuable suppliers. This contributes to strengthen their established socio-economic infrastructures and relationships, and may well even expand these capabilities by complementing new technological opportunities with existing accumulated knowledge. Supplying additional parts or entering new relationships with other industries (e.g. the application of new materials technologies developed in aerospace to automobiles) will expand the competences and capabilities of these clusters (see the contributions by Greef and Schroeder, Baker et al. and Scherrer; Huber 2012). Thus, multipurpose technologies, such as Industry 4.0 and digitization, provide opportunities both for constant innovations employing established structures and using new technological opportunities for additional networking and building new supply chains. Since there are many specialized competences and an increasing global division of labour opportunities are created where both R&D and manufacturing can be advanced to build new networks and supply chains (see the contribution by Hickie and Hickie). What is specific about these technologies is the wide range of applications and exploitation that can achieved in collaboration with existing expertise and organizations from different areas. This allows for some reorganization of value chains and, while managing this, it demands a particular understanding of manufacturing processes in the light of costs, quality, transportation and reliability, knowledge of which already exists based on experience.

Building such new centres of expertise and organizing clusters of smart factories are both achieved in situations which are highly integrated into global supply chains and interdependent with regard to innovative processes in manufacturing and technology development. Thus, transportation infrastructure as well as state-of-the-art data transmission is a necessary requirement. Creating new locations to participate and new centres to contribute needs a close relationship with metropolitan areas and the strengths provided there (Hilpert 2018). Additional, new locations are confronted with a number of constraints. Their location needs to provide opportunities to make their own particular contributions and they need appropriate transportation links. So long as the established regions or metropolises continue to take and cope with new technological opportunities the room for new locations is rather limited. Despite the changes made possible by Industry 4.0 and digitization, still, established locations can organize their resources to enable economically sustainable development. Where such new technologies are being developed and created challenges are faced from competitors reflecting both their abilities to collaborate and to contribute (Hickie 2006). In established regions, this competence merges with their existing competences from their earlier experiences and application of new technologies it also enables a continuation of existing or improved product lines.

Consequently, Industry 4.0 and digitization indicate the potential for sustainability alongside new opportunities at other locations. The situation becomes more complex, but it clearly indicates an industry-location-nexus that empirically is expressed in highly divergent ways. There are simultaneous opportunities for continuity and change if technological innovation can be managed and organized.

Windows of innovation facing divergent situations

The close relationship between Industry 4.0 and digitization with existing structures of regions and metropolises also inhibits rather divergent levels of preparation or readiness for the areas of innovation (Florida, Adler, and Mellander 2017). Manufacturing industries respond differently to opportunities of such new technologies. The contributions of parts of different complexities play an important role when building supply chains. While in consumer products (e.g. consumer electronics, household goods or textiles and clothing), many simple parts are contributed, there are more complex elements when it comes to products of higher investments or complex equipment (e.g. automobiles, aircraft, mechanical engineering, measurement instruments or medical instruments) such opportunities are rather limited and particular attention is paid on the quality and reliability of supply. Besides, advanced products of life sciences, pharmaceuticals or specialized chemicals are differently organized and refer to specific areas where to apply such new technological opportunities. Considering the different levels of innovation with regard to the parts contributed to supply chains, consequently, situations are highly divergent (see the contribution by Hickie and Hickie).

Such divergent opportunities also indicate that technologies which apply for certain stages of manufacturing may not apply for others. Given the wide range of opportunities associated with Industry 4.0 and digitization technologies and the situation of application are decisive for innovative processes. As a consequence, there are different windows of innovation which will become available for divergent situations. Clearly, innovations which demand for highly skilled and educated labour will not be available for regions, where the labour force does not meet the requirements (Vassiliadis 2019; Vassiliadis and Hilpert 2020; and see the contributions by Greef and Schroeder, Hickie and Hickie, Gaddi et al). Their window of innovation demands for technologies to be applied under the condition of less skilled workforce. When improving the level of skills and education (Chung 2019), the situation changes towards higher values added and more complex processes of innovation. But, due to differences in labour costs and regulations, those applications which can be realized with less skilled labour will not be available anymore. Consequently, windows of innovation will open according to the match of situation and technology and whether the region may change for the level of innovation which suits (Gaddi, Garibaldo, and Garbellini 2020).

While this match is defined also by the changes during processes of social and economic development, there is also impact by technologies such as 3D printers or change of manufacturing which may replace even cheap labour and weak regulations even when considering both the costs of transportation and the closeness to markets (see the contribution by Hickie and Hickie). Thus, supply chains may also change, when the window of innovation allows for re-localization of manufacturing based on new technologies despite higher labour costs. Regions are facing constantly changing situations concerning their structures and the networks they are contributing to. Technologies developed at a region and related with a transregional network across regions, countries or even continents are familiar to those regions because frequently they continue an ongoing development by advancing existing capabilities. Other regions, which are not related with such technological developments, are confronted with these technologies newly, there such technologies are applied as something almost entirely new which provides the impression of disruptive change (see the contribution by Scherrer; O'Gorman and Donelly 2020). While participating in these innovative processes and manufacturing based on this new equipment situations are constantly changing, but even regions with particularly strong industrial structures (see the contributions by Sandulli et al., Gaddi et al., Greef and Schroeder) get used to the changes in technology, organization and management.

Consequently, there are divergent processes of innovation according to the application of technologies or whether they provide the potential to develop and manufacture what is considered to be the next-generation equipment. Both the perception and the responses, of course, are different. These processes are facing changing contexts because of both the adoption of technologies and of the industries the enterprises belong to. In addition, competition, skilled and educated labour, and potentially new suppliers are forming a global context of ongoing changes and dynamisms. Thus, based on different technologies and situations of industries or along the supply chains, there are divergent situations characterized by windows of opportunities which are opening and closing according to the processes they are related with (see the contribution by Scherrer). These constantly changing contexts also are characterizing the challenges and difficulties of adjusting industrial strategies (see Sandulli et al.). During the time of such processes, there is also demand for a constant modification or even wider adjustment of strategies. Even new strategies can be designed either in addition to existing ones or replacing those.

Thus, Industry 4.0 and digitization are not just confronting regions and metropolises with new situations which they are placed in, but, in addition, new windows of opportunities are required. The wide range of technologies means that there are divergent opportunities which need to be identified and used, but this needs to be realized within the time frame which is defined by the individual window of opportunity. Consequently, there is a continuum of change associated with time and situation, which, in addition, is related to international contexts of markets and supply chains. This is even more the case, if these are value chains associated with individual contributions to the innovative capability of the parts because this provides for a synergy that builds up by contributing to an even more innovative final product of the OEM (see the contribution by Hickie and Hickie). Consequently, some technologically new opportunities suit with some parts and can create important effects in a number of regions or metropolises where such parts are manufactured to be contributed to such an innovative supply chain. Similarly, other new opportunities may suit predominantly elsewhere and at different points in time.

There is a similar impact of digitization which can be identified with services. Contributions in accounting, print media and Internet presentations or specific pieces of work, which can be announced for offers on platforms, are already well known and did take influence on where such work shall be done. Still, when it comes to software engineering, it can be noticed that contracts are given to people who qualify for this but who live in low-cost countries and, consequently, are in a position to make offers more attractive as labour is cheaper. More recent development in digitization are controlling machines, aircraft, automobiles etc. while these are in use. While before services were provided regularly after the equipment was used for a certain time or distance today, these are prepared to announce on duty that there is a need for services or material to be replaced or adjusted (see the contribution by Hickie and Hickie). In addition, increasingly such services required can be done remotely or by robots which are steered from far away (see the contribution by Gaddi et al.). Based on opportunities of digitization, also modern medical devices can be used even for the benefits of the patients from experts being geographically far away.

Although the digitalized technologies allow for the use of distant competences, it does not change the concentration of competences at selected regions or metropolises (see the contribution by Hilpert). In contrast, these centres of competences become particularly important as these will help to improve services in other regions (Ciffolilli and Muscio 2018). This reduces the need for the availability of all competences and capabilities at a particular place. Nevertheless, the demand for advanced digital infrastructure at both ends will induce a concentration of competences at certain places, because there are those from where the services are provided and related to capabilities and equipment at other locations which receive the support (Götz and Jankowska 2017). Similar to supply chains of industrial manufacturing, there are tendencies for advanced service supply chains (see the contribution by Baker et al.). Consequently, due to digitization also in services, there are windows of innovation which can be used at regions and metropolises where infrastructure and educated labour suit to provide the services required. Consequently, activities may become organized globally by supply chains and value chains and allow for a participation in emerging economies, but again also in these countries, there are strong tendencies of concentration in metropolises as these supply infrastructures necessary (Chen et al. 2019).

Services to be transferred through the Internet widely demand for appropriate skills which are frequently based on university education. This will allow us to supply particular services which easily can be implemented in processes elsewhere while contributing from locations around the world. Whereas the contribution of educated labour which demands a close collaboration which frequently is realized in shared laboratories or other equipment of R&D will rarely be realized virtually being based on digital networks. While the window of opportunities can open for such locations in emerging economies as far as Taylorized contributions can be supplied, in contrast services which demand for both complex *in situ* collaboration and outstanding competences create a strong brain drain to established centres which are globally well known. Also, in innovation related to digital technologies, income and working conditions continue to be effective when organizing situations which allow us to change towards more attractive arrangements, as well as contribute to the chains with personnel that will stay at the regions or metropolises (Abel and Deitz 2009; Hilpert 2014; Sandulli and Giménez Fernández 2019).

This also highlights the role of societal structures to realize innovation processes. Regional societies with a high percentage of highly skilled blue-collar labour and those with university degrees provide a basis for economies which are oriented in innovation and high value-added products. This refers to an appropriate labour force as well as to enterprises which engage in innovative activities and developments. Dynamic attitudes among entrepreneurs and start-ups are also important to create a labour market for attracts innovative employees and to relate benefits with a region (see the contribution by Gaddi et al. and Scherrer). Societal structures and interest groups representing entrepreneurs, unions and such skilled and educated groups are forming the structures which are fundamental. Consequently, there are diverse regional societal structures which correspond with regions and metropolises which makes it hard for some regions to meet the requirements concerning enterprises and labour available.

Consequently, organizational structures of labour costs still are important when organizational structures of the value chains are favouring certain locations. The windows of opportunities open for regions and metropolises, and those in Europe and North America can take advantage of such chains of collaboration when the mix of competences and opportunities suits to complement their opportunities. Thus, the divergencies of innovative opportunities are both characterized by the individual structures and competences and by the context which may provide matching opportunities regarding manufacturing or services (Baker, Drev, and Almeida 2019; Hickie, Jones, and Schloderer 2019). Nevertheless, also in times of Industry 4.0 and digitization when skills and competences are emphasized, still labour costs and regulations matter. But, when infrastructures become critical even this situation is characterized by a threshold to qualify for participation and contribution, because there needs to be a match with Internet, modern communication and, last not least, transportation of goods and human capital.

The new opportunities as well as the reorganization associated with windows of innovation by Industry 4.0 and digitization are rather selective concerning the regions and metropolises which may take advantage. Since the windows of innovation associated with these processes refer to the global chains, it will include a greater number of locations relating emerging economies with European and North American development, but it will not introduce a systematic de-concentration rather it will induce a technology-related multipolarity. Driven by the new technologies, the complex interrelationship with reorganization and management of chains and processes again emphasizes the importance of divergent competences and capabilities. The application of such new technological opportunities as well as R&D on digitalized technologies and its manufacturing closely relates to particular personnel which suits structures and context of innovative change. These situations are systematically diverging and can meet individual windows of innovation only when matching competences can be generated or transferred to the regions or metropolises which qualify.

Government policies in the light of regional capabilities: continuation of uneven development

Constantly changing international contexts and new technological opportunities are fundamental for regional and metropolitan situations and how these might adopt to new situations (see Hilpert). The opportunities are different, when it is a simple supply chain or when value chains are addressed. Value chains are characterized by the continuously increasing value of the final products. Individual participants in these chains are contributing to the final innovative level of the products or services (see the contributions by Hickie and Hickie, Baker et al.). When it comes to less complex products which are providing the basis for supply chains, a certain quality can be realized at many locations under the guidance and supervision of the OEMs which are assembling or merging such parts as marketable products. Although opportunities of Industry 4.0 and digital technologies are important in both situations, the processes differ fundamentally. While innovative processes within value chains help contributing to generate higher values added, in contrast, in supply chains, such improved equipment is used to reduce costs and to manufacture more efficiently.

Consequently, regional and metropolitan situations are constantly changing but asking for divergent responses and capabilities (see the contribution by Hilpert). Since industrial structures and R&D orientations vary according to technologies which might suit the situations, long waves of technologies will have a divergent regional impact that will change during the processes of introduction as well as of maturation (see the contribution by Scherrer; Scherrer 2020). Each phase of these waves is based on the further development of existing technologies or on the application of opportunities across technologies and industrial sectors and will demand for specific responses by building capabilities and competences at locations related with value chains or supply chains (see the contributions by Baker et al., Gaddi et al., Hickie and Hickie). In particular, at locations of higher competences in R&D, there will be opportunities of cross-fertilization or 'wave changes' based on merging competences of different origin. As lighter materials will influence aircraft and automobile industries, digitization can help changing transportation or making energy consumption more efficient and, last not least, based on Industry 4.0, the steering of machines or instruments can be realized more precisely.

During the time of technological development, reorganization and adjusting management of the chains, simultaneously, there is constant change at individual locations concerning structures of research, industries and capabilities of skills and education. For some locations related to chains, this allows us to improve their capabilities and to become placed on the chains based on more attractive contributions, while others may prepare for a continuation of their position on the chain by arranging for such changes induced by Industry 4.0 and digitization (see the contributions by Gaddi et al., Greef and Schroeder). The contribution to these chains is widely based on the capabilities which are available in the regions or metropolises. The development of educational systems to support individual skills and building research structures which complement the industrial capabilities or services may help to continue and improve in the light of Industry 4.0 and new digital technologies. The strong impact of these new technologies and the organization of the chains clearly emphasize the importance of skills, education and research.

When arranging for such processes, consequently, public policies on innovation, research and education become very important. National and regional programmes to support for Industry 4.0 and digitization prepare for the development or adaptation of technologies (see the contributions by Sandulli et al., Greef and Schroeder, Gaddi et al., Baker et al.). Based on these situations, existing industries can either continue their product development and research strategies based on their competences and in relation with universities and research institutes, or industries of the regions in question can arrange for the application of such new technological equipment. Although structures are widely continued, they are continued on a fresh basis. Both established value chains and supply chains form important conditions for the way such new technologies are linked with regions. When each level of supplier contributes innovative value into the parts to form a better final product, of course, both are involved the application of new technologies as well as an improved design of the product. On the other hand, when the products remain widely unchanged, the new technologies will be introduced to produce better, improve the reliability of quality and at lower costs.

While these new technologies allow for such changes within in the chains considered or among the regions and metropolises included, still, strategies of OEMs play an important role. They need to be considered when public policies are designed and when there are ideas about individual development which may address new opportunities based on other firms or start-ups in particular areas of competences. New competences brought to the region along with Industry 4.0 and digitization may also be applied in other industries or with other economic activities. Traditional industries may take advantage of manufacturing when assisted by 3D printers which are used on the basis of designs transmitted via Internet. Transportation over short distances allow for being close to markets, new products based on traditional competences may refer to cross-sectoral fertilization when adopted in new situations and may make even those industries to be innovative which were about to face processes of maturation (see the contribution by Scherrer). Thus, competences associated with Industry 4.0 and digitization may become effective even beyond the original area of R&D or application.

Equipment by Internet access may allow for smart factories and remote steering, but it may also provide the basis for certain advanced services given appropriately educated personnel is available (Carvalho et al. 2018). Consequently, a new infrastructure required for these new technologies may also provide opportunities of structural change or modernization and may help to open new areas of economic activity and employment (Ibarra, Ganzarain, and Igarta 2018). This indicates the particular contribution of public policies with regard to technologies which reach out so far and are as multi-conditional as the introduction of Industry 4.0 and digitization. The infrastructure required and provided publicly allows for communication, exchange of information and communication beyond the industries which were addressed. The availability of personnel ready to apply these technologies will help to take advantage of new opportunities and educational systems supplying both skills and degree holders to work with these new technologies indicate the close linkage between human capital and socio-economic development (see contributions by Baker et al. and Gaddi et al.; OECD 2017). Public policies providing for such labour force and supporting appropriate research become critical for future development (Dolphin 2015; O'Sullivan and Mitchell 2013).

Technologies of Industry 4.0 and digitization thus indicate an increasing interdependence between public policies and socio-economic development. Although this is not a disruptive innovation, it is widely fundamental for future development. Consequently, governments on different levels execute programmes and initiatives of divergent focuses. Regional governments of strong budgets design programmes which suit the industrial structures and support research capabilities (see the contributions by Sandulli et al., Greef and Schroeder). German Länder gives additional attention to competences, skills and research which either suit to support the application of new digital technologies while improving industrial equipment or to develop new technologies. Länder like Bavaria, Baden-Württemberg or Northrhine-Westphalia have large budgets which allowed us to design programmes addressing enormous amounts (e.g. Northrhine-Westphalia invests € 450 millions), in contrast, Länder of smaller budgets struggle not to be left behind (e.g. Brandenburg, Bremen, Lower Saxony or Saarland). Consequently, industrially strong and innovative Länder will be restrengthened due to innovative progress along with Industry 4.0 and digitization. In addition, national research funds are allocated where the strongest capabilities and expertise are, and also supporting these existing and well-developed centres of excellence.

In countries like Italy which are not federally organized and thus regions have smaller budgets national governments are considered to be particularly important. The funding is addressed to enterprises and research capabilities which suit best for Industry 4.0 and digitization. Due to clustering and developed industrial structures, this concentrates on particular regions in Northern Italy (see the contribution by Gaddi et al.; Castelo-Branco, Cruz-Jesus, and Oliveira 2019; Götz and Jankowska 2017). The government support is addressed to enterprises which are predominantly suppliers to German and French OEMs. Established contacts and the situation as reliable supplier provide for socio-economic development which is stronger than in other regions of the country. Because in the South, there is a lack of enterprises and locations to match the requirements; consequently, these disadvantaged regions lack enterprises which would be ready to contribute to advanced value chains. The lacking competences to join the chains clearly create disadvantages for the process of innovation because these are based on existing structures. Thus, industrial competences, products and skilled personnel are lacking, which are required to participate in reorganization and management of innovation. The arrangements required for the application of these new technologies are not available. An even increasing interregional grading is to be awaited.

Existing or potential industrial structures and product capabilities are clearly important for participation in innovation based on Industry 4.0 and digitization (Castelo-Branco, Cruz-Jesus, and Oliveira 2019; Ibarra, Ganzarain, and Igarta 2018). Similarly, in Spain, regions of advanced structures such as Catalonia and the Basque Country provide structures that could be prepared to enjoy new technologies. But Spanish suppliers frequently are manufacturing plants within the country providing products of less complexity. On the other hand, the region of Madrid is the home of a large number of software engineers and start-up enterprises based on this competence (see the contribution by Sandulli et al.). This situation draws international attention as costs may be lower than in other European countries and government funding to prepare for digitization will find appropriate recipients predominantly in metropolitan regions (see the contribution by Baker et al.). Areas which do not provide for structures of industries and skilled labour which may suit for such new technologies may meet difficulties to benefit from such opportunities. Although there are regions which characterized as locations of suppliers to OEMs elsewhere or depending on new machines and equipment from abroad these regions develop differently from others within a country and when compared within the country they often continue becoming even stronger. Simultaneously, there is a tendency of skilled and educated workforce to migrate to such regions and metropolises (Giménez Fernández and Sandulli 2020), thus restrengthen the societal basis for innovative development.

Government programmes are addressed to existing situations and structures and aim at strengthening opportunities of industries or services. The increasing development of both supply chains and value chains provides for highly diverse situations of the regions and metropolises. Public policies, of course, support the opportunities of a region or country and, thus, frequently existing relationships and opportunities of development are guiding when such policies are designed. Policies find highly diverse situations and respond by appropriately and diversely designed policies. The wide range of applications associated with Industry 4.0 and digitization also indicates diverse areas of public support, although there is a generally converging focus on these technologies. Strength in research and industry or high-quality services are addressed and helps to continue or even advance socio-economic development and the position within chains. Simultaneously, such structures and capabilities also inhibit opportunities to develop new areas of activities, when new and matching technologies become available. Thus, the problems of regions which are lacking such capable structures and linkages to supply chains or value chains will continue having difficulties in their future socio-economic development. In contrast, those which are at least supplying valuable parts perform as the recipient of such policies and can take advantage of such change.

Conclusions: innovative processes and societal tendencies of new technological opportunities

For many enterprises and employees, digital technologies and Industry 4.0 provide the impression of a highly unexpected process that induces fundamental changes. Regions and metropolises are facing changes associated with such processes of innovation as a sudden change which in their particular situation even may have some similarities with disruptive innovation. Based on both new equipment and Internet infrastructure, indeed, traditional manufacturing and services provided may not be continued. In contrast, in regions and metropolises where such technologies are developed, this is a continuation of both research strategies and competences in the design of high-tech products. In such Islands of Innovation frequently, there are several competences of technologies and product development which can take advantage based on collaboration across these different capabilities. This also indicates their particular contributions to value chains and supply chains by enterprises or even clusters at different locations (Ferraro and Iovanella 2017). There is an increasing tendency for a division of innovation when it comes to joint R&D but also along the chains by innovation which is contributed within the parts delivered or modern services provided. Digitization allows for such processes and Industry 4.0 provides the basis for reorganization and management of manufacturing such products.

This clearly creates effects towards selective networking. Research capabilities, enterprises and locations can participate in such collaboration if skills, educated personnel, research capabilities, industrial structures and fast transportation is provided. Whereas the extent to which such elements are required and arranged is highly diverse. Consequently, there is a common impact of such new technologies but, despite the fact due to these new technological opportunities, the individual situations are widely different. Thus, according to the kind of contributions, there is a simultaneity of concentration on particular regions and a polycentric de-concentration towards a number of regions or metropolises within the networks (Carvalho et al. 2018; Götz and Jankowska 2017). While a wider range of participation in such networks can be identified when it comes to manufacturing and services as well as opportunities of relocation of some manufacturing due to a revaluation of labour costs, the participation is selective and is associated with societal tendencies for both enabling to participate and the kind of contribution to the networks.

Consequently, divergencies in situations, industrial structures and capabilities create different demands for labour and are related to divergent composition of the labour forces of both regions and metropolises contributing to the networks. Thus, although these locations are interlinked through networks and chains, the regional societies are divergent. While in some cases, there is a stronger demand for university-educated personnel to be engaged in R&D, others are in need for more skilled blue-collar labour to realize manufacturing. Consequently, these are divergent situations which converge with structures of the regional societies and their innovative potential based on the capabilities of the labour force. In particular, value chains or the division of innovation refer to societal constitutions which allow for higher values added. In addition, public education and strong research capabilities in areas which suit these arrangements contribute to networking and chains. The complementarities of the societal situations are decisive for the realization of socio-economic development within the networks and based on the different chains.

The way the development and application of technologies of Industry 4.0 and digitization indicate tendencies towards spatial patterns of societal divergencies which are indicated by regions and metropolises participating based on the potential of the labour forces which is provided within the networks and chains. Both, regions and metropolises, are associated with these new technologies, the exclusive conditions to participate within the networks and chains, and the individual level of development associated with the position within these contexts. In particular, smart factories refer to trans-regionalization because the steering and management to a far extent can be realized remotely, which reduces many locations and plants to execute ideas developed abroad. Nevertheless, entering value chains by building capabilities for advanced products of higher values added would allow us to regain independence by providing innovative parts of high quality which would make regions and metropolises to become the home of firms which relevantly contribute to final products.

Industry 4.0 and digitization indicates how the context of innovation changes towards centralization of decision taking, dependencies of plants and locations as well as opportunities can be created based on policies to change from a supplier to a contributor to innovation. The more regions and metropolises can manage to build such innovative competences and a corresponding innovative labour force, the more they can participate in innovative process and products. Consequently, many changes may not appear as disruptive innovations but as a continuation of R&D. This also indicates that innovative regional societies and their capabilities of labour help to improve the receptiveness of new technologies, although the ways these are realized are rather diverse.

Disclosure statement

No potential conflict of interest was reported by the author(s).

References

Abel, J., and R. Deitz. 2009. "Do Colleges and Universities Increase Their Regions Human Capital?" *Federal Reserve Bank of New York: Staff Reports No. 401.*

Baker, P., M. Drev, and M. Almeida. 2019. "Postsecondary Education and the Development of Skilled Workforces: Comparative Policy Innovation in Brazil and the U.S." In *Diversities of Innovation*, edited by U. Hilpert, 108–136. London: Routledge Ltd.

Bellini, N., and C. Pasquinelli. 2018. "Branding the Innovation Place: Managing the Soft Infrastructure of Innovation." In *Handbook of Politics and Technology*, edited by U. Hilpert, 79–90. London and New York: Routledge Ltd. (paperback).

Carvalho, N., O. Chaim, E. Cazarini, and M. Gerolamo. 2018. "Manufacturing in the Fourth Industrial Revolution: A Positive Prospect in Sustaining Manufacturing." *Procedia Manufacturing* 21: 671–678. doi:10.1016/j.promfg.2018.02.170

Castelo-Branco, I., F. Cruz-Jesus, and T. Oliveira. 2019. "Assessing Industry 4.0 Readiness in Manufacturing: Evidence for the European Union." *Computers in Industry* 107: 22–32. doi:10.1016/j.compind.2019.01.007

Chen, X., R. Li, X. Niu, and V. Hundstock. 2019. "Diversified Metropolitan Innovation in China." In *Diversities of Innovation*, edited by U. Hilpert, 210–236. London: Routledge Ltd.

Chung, S. 2019. "South Korea as a New Player in Global Innovation: Role of Highly Educated Labour Force's Participation in new Technologies and Industries." In *Diversities of Innovation*, edited by U. Hilpert, 179–209. London: Routledge Ltd.

Ciffolilli, A., and A. Muscio. 2018. "Industry 4.0: National and Regional Comparative Advantages in Key Enabling Technologies." *European Planning Studies* 26 (12): 2323–2343. doi:10.1080/09654313.2018.1529145

Dolphin, T. 2015. *Technology, Globalisation and the Future of Work in Europe. Essays on Employment in a Digitalised Economy.* London: Institute for Public Policy research (IPPR), 119 p.

Ferraro, G., and A. Iovanella. 2017. "Technology Transfer in Innovation Networks: An Empirical Study of the Enterprise Europe Network." *International Journal of Engineering Business Management* 9: 1–14. doi:10.1177/1847979017735748

Florida, R., P. Adler, and C. Mellander. 2017. "The City as Innovation Machine." *Regional Studies* 51 (1): 86–96. doi:10.1080/00343404.2016.1255324

Gaddi, M., F. Garibaldo, and N. Garbellini. 2020. "The Italian Experience in Implementing Industry 4.0." *UCJC Business and Society Review* 17 (2): 52–69.

Giménez Fernández, E., and F. Sandulli. 2020. "High Skilled Workers Mobility in Spain and Europe: Motivations and Implications." *UCJC Business and Society Review* 17 (2): 70–89.

Götz, M., and B. Jankowska. 2017. "Clusters and Industry 4.0–Do They Fit Together?" *European Planning Studies (EPS)* 25 (9): 1633–1653. doi:10.1080/09654313.2017.1327037

Hickie, D. 2006. "Knowledge and Competitiveness in the Aerospace Industry: The Cases of Toulouse, Seattle and North-West England." *European Planning Studies (EPS)* 14 (5): 697–716. doi:10.1080/09654310500500254

Hickie, D. 2020. "Diversities of Innovation in Advanced Manufacturing – the Aerospace and Automotive Industries." *UCJC Business and Society Review* 17 (2): 50–56.

Hickie, D., N. Jones, and F. Schloderer. 2019. "Globalisation, Competitiveness and the Supply of Highly Skilled Labour in Civil Aerospace." In *Diversities of Innovation*, edited by U. Hilpert, 275–295. London: Routledge Ltd.

Hilpert, U. 2014. "Networking Innovative Regional Labour Markets. Towards Spatial Concentration and Mutual Exchange of Competence, Knowledge and Synergy." In *Networking Regionalised Innovative Labour Markets*, edited by U. Hilpert, and H. Lawton Smith, 3–31. London: Routledge Ltd.

Hilpert, U. 2018. "The Culture-technology Nexus: Innovation, Policy and Successful Metropolis." In *Handbook of Politics and Technology*, edited by U. Hilpert, 149–161. London and New York: Routledge Ltd.

Hilpert, U., and H. Lawton Smith, eds. 2014. *Networking Regionalised Innovative Labour Markets*. London and New York: Routledge Ltd. xvii+204p. (paperback).

Huber, F. 2012. "Do Clusters Really Matter for Innovation Practices in Information Technology? Questioning the Significance of Technological Knowledge Spillovers." *Journal of Economic Geography* 12: 107–126. doi:10.1093/jeg/lbq058

Ibarra, D., J. Ganzarain, and J. I. Igarta. 2018. "Business Model Innovation Through Industry 4.0: A Review." *Procedia Manufacturing* 22: 4–10. doi:10.1016/j.promfg.2018.03.002

Kerr, S. P., W. Kerr, Ç Özden, and C. Parsons. 2017. "High-skilled Migration and Agglomeration." *Annual Review of Economics* 9: 201–234. doi:10.1146/annurev-economics-063016-103705

Lawton Smith, H., and D. Assimakopoulos. 2020. "'Islands of Innovation' and Diversities of Innovation in the UK and France." *UCJC Business and Society Review* 17 (2): 18–35.

OECD. 2017. *OECD Employment Outlook 2017*. Paris: OECD Publishing, 216 p.

O'Gorman, W., and W. Donelly. 2020. "eInnovation: The Importance of Context." *UCJC Business and Society Review* 17 (2): 36–51.

O'Sullivan, E., and N. Mitchell. 2013. "International Approaches to Understand the Future of Manufacturing." *Future of Manufacturiong project, No. 26, Government Office for Science*, London.

Sandulli, F., and E. Giménez Fernández. 2019. "Underemployment of Middle-skilled Workers and Innovation Outcomes: A Cross-country Study." In *Diversities of Innovation*, edited by U. Hilpert, 137–153. London: Routledge Ltd.

Scherrer, W. 2020. "How 'General Purpose Technologies' Trigger Long Waves of Economic Development and Thereby Generate Diversities of Innovation." *UCJC Business and Society Review* 17 (1): 36–49.

Vassiliadis, M. 2019. "Skilled Labour and Continuing Education: The Role of Social Partners for Divergent Opportunities of Innovation." In *Diversities of Innovation*, edited by U. Hilpert, 35–48. London: Routledge Ltd.

Vassiliadis, M., and Y. Hilpert. 2020. "Labor-based Innovation: The Advantage of Skills and Education." *UCJC Business and Society Review* 17 (2): 66–83.

The impact of Industry 4.0 on supply chains and regions: innovation in the aerospace and automotive industries

Desmond Hickie and James Hickie

ABSTRACT

This paper explains the spread of Industry 4.0 technologies in two globalized, manufacturing industries. It then goes on to suggest ways in which the widespread adoption and integration of these technologies may impact upon their geographical distribution – not least engaging new regions and perhaps altering the significance of established ones.

Introduction

Automotive and civil aerospace have much in common. Each has a complex and highly stratified, globalized supply chain. Significantly, both industries are already widely engaged in the adoption of Industry 4.0 technologies. The two industries are both globalized and regionalized – on specialized islands of innovation with long industrial histories, like Detroit in automotive and Toulouse in aerospace. These regional[1] agglomerations, of automotive or aerospace businesses, also contain other sectorally-focused resources (e.g. universities, research institutes) and, most importantly, have a capacity to innovate (Hickie, Jones, and Schloderer 2019; Hilpert 2019). There are technological spillovers between the sectors, for example from automotive to aerospace in robotics and from aerospace to automotive in composite materials. The key structural distinction between the sectors is that civil aerospace is a Boeing-Airbus duopoly, whilst automotive Original Equipment Manufacturers (OEMs) are competing in an increasingly overcrowded marketplace. Industry 4.0 technologies are already bringing significant changes to the structure of supply chains and their global distribution in both sectors. The case made here is that the increasing adoption of digital technologies is likely to further enhance these tendencies in the coming decade.

The character and impact of Industry 4.0

Whilst the term Industry 4.0 has been used in various contexts and its boundaries appear somewhat flexible (e.g. Brettel et al. 2014), it is clear that it is driven by the rapid growth in our ability to gather and process data, and then to transmit and store it. More particularly, although Artificial Intelligence (AI) has been around for decades (Hickie 1991; Hilpert 1991), recent, rapid developments in AI have been driven by greatly enhanced computer power, big data and advances in machine learning (Rotman 2017) and by the falling costs of new technologies. Although the automotive and aerospace industries have been major users of computers since the creation of the silicon chip in 1973 (e.g. in robotics, Baldwin 2016), rapid advances in Industry 4.0 technologies offer major opportunities for rapid innovation.

Industry 4.0 technologies create opportunities for enhanced synergies and greater optimization of activities across the board, from R&D to customer relations. For example, Augmented Reality (AR) can allow access for manufacturing workers or consumers to expert advice, without an expert present. Virtualizing supply chains can enhance inter-company collaborations and the closer integration of companies' core competences (Brettel et al. 2014). Telemigration can allow firms to draw on the expertize of part-time or contract workers to provide a cheaper, more flexible workforce, a prospect that will grow with improvements in machine translation and telepresence technologies (Baldwin 2019). Advances in the digital economy have driven a move from capital-based value creation to knowledge-based value creation and, in advanced economies, this has led to a shift in the balance of investments away from investment in capital goods and towards investments in intangibles (e.g. R&D, design, software, Haskell and Westlake 2017).

The advance of the digital economy is likely to have profound effects on employment. Beginning in the 1970s the ICT revolution saw the displacement of skilled and semi-skilled manual workers' jobs by robots. However, whilst this was highly detrimental for certain types of worker, its effects on employment were not always detrimental. Many highly skilled, well paid service jobs were created for those who could write software, manage robots and so on. Industry 4.0 technologies are likely to have a similarly disruptive effect on labour markets. As a World Economic Forum (2016) report suggested, automation will displace many current jobs and even occupations. However, in this case, the jobs displaced are more likely to be service jobs, in particular those performed by less well-educated employees (Rotman 2017). Those who continue in employment are likely to find that they need to acquire new knowledge and skills to compete effectively in the labour market (Wright 2019). For example, in the 'smart factory' workers are likely to become coordinators of machines and problem solvers when things go wrong, whilst managers will need to become more skilled at working with information (Grybowska and Lupicka 2017). This is not to say that computers will perform better than humans at all tasks, or even be able to undertake all tasks. Currently, humans are better than machines at understanding language, although increased computer power and larger databases may reduce aspects of that advantage (McKinsey Global Institute 2017). However, humans have qualities which give them 'unique advantages' (Baldwin 2019, 236), like using judgement when faced with incomplete data, that are not replicable by machines. Similarly, humans can explain their decisions and take responsibility for them, both matters of great significance in automotive and aerospace.

The specific impacts of Industry 4.0 technologies on aerospace and automotive are too recent to be well represented in the academic literature. There are studies of digitalization which helpfully use these sectors as examples (e.g. Grendel et al. (2017) point out that digital technologies are more immediately useful in highly automated sectors (like automotive) rather than in sectors which involve significant levels of manual assembly to create very large structures (like aerospace); see also Theorin et al. (2017)). Similarly, Ceruti et al. (2019) look at the potential uses of additive manufacturing and AR in aerospace MRO. Other writers report specific technology projects to support one sector or the other (e.g. Valachek et al. (2017) describe a project to construct digital twins to support the Slovak automotive industry). Beyond this there are non-academic reports: with a general digital technology focus (e.g. the World Economic Forum's (2016) study of the potential impacts of digitalization on employment; or, PwC's (2018) study of digital champions in the Asia-Pacific region); and, reports with a specifically sectoral focus (e.g. KPMG's analysis of mobility in 2030 (2019b)). The business focused press (e.g. The Financial Times, The Economist) and the professional and sectoral press (e.g. The Engineer, Aerospace) also contain useful and up-to-date insights into both sectors. Finally, some major consultancies are producing advice about how these sectors might respond to the Covid crisis using digital technologies (e.g. PwC (2020a, 2020b); Times Tempus (2020)). Inevitably, given the rapid advances in digitalization, much of what is written for current consumption is inevitably speculative.

Some impacts of digitalization upon the geographical distribution of the aerospace and automotive industries can be discerned, but are presently still in their infancy. What is well established is that innovation in both industries is already highly regionalized in locations that have strong, specialized regional innovative ecosystems (Hickie 2006; Adner 2017) which support Research and Development (R&D)and new product and process developments, for example in universities and through networks of experienced specialist suppliers (Hickie, Jones, and Schloderer 2019). Furthermore, these specialized regions can be understood as islands of innovation (Hilpert 2019), not innovating in isolation, but collaborating with other geographically dispersed technologically advanced regions or islands (Hilpert 1991). A similar process of regional concentration can be found in digital technologies, most obviously in Silicon Valley. Clustering brings with it various benefits, which both keep established firms in a region and attract new firms to it, and such clustering tends to be more common for knowledge-based R&D activities than for production (Alcacer and Chung 2007; Siedsschlag et al. 2013).

The potential impacts and challenges of Industry 4.0 technologies

Technological innovation is critical to competitiveness in both industries. Significant advances have taken place, for example in the use of automation, composites, driver aids and fly-by-wire technologies. However, it is arguable that the technological trajectory of both sectors has been incremental since the advent of jet airliners sixty years ago. Industry 4.0 technologies have the potential to underpin more radical changes in both sectors, significantly increasing the pace of both product and process innovations. In both sectors Industry 4.0 technologies have the innovative potential to engage both mature, well established firms and new entrants with niche expertize, and to impact the regional distribution of their innovative activities.

Managers in both sectors see Industry 4.0 technologies creating a step change in product innovations (e.g. KPMG. 2019a). The most critical are in vehicle autonomy, shared passenger mobility, product deliveries and the enhancement of customer relationships. In automobiles, advances have already allowed vehicles to take over functions from drivers, and are leading towards either to semi- or wholly autonomous vehicles (http://

tech.crunch.com 2016). It is clear that global digital technology companies (e.g. Uber and Waymo – an Alphabet subsidiary) are undertaking ambitious research to develop fully autonomous vehicles. Autonomous vehicles create the possibility for new services with far reaching consequences (Economist 2018). Digitization and autonomous vehicles have the potential to reshape individual mobility, allowing on-demand consumer mobility and using autonomous vehicles which can radically cut consumer costs. Greater connectivity also allows for the development of Mobility as a Service integration, in which timetabling and payment for different forms of transport can be integrated. Over the next decade or so, these developments may reduce the need for individual and company car ownership (PwC 2019). In the automotive sector they may also radically change the relationship between manufacturers and customers, for example enabling manufacturers to establish individual relationships with customers bypassing dealerships (PwC 2020b). Immediate ambitions for large, autonomous airliners are less advanced, reflecting passengers' and regulators' wariness. However, both sectors are involved in developing new, faster, cheaper delivery ecosystems, radically transforming retailing. In aerospace there are drones and cargo planes with only a single pilot under development to speed up short- to medium-range deliveries. These can complement self-driving delivery platforms (e.g. the Toyota e-Palette), localized delivery robots and dynamic routing systems (e.g. the Via Van – a Via and Mercedes joint venture) (KPMG 2018).

Industry 4.0 technologies are key to speeding up design processes to bring new vehicles to market more quickly. For example, in automotive it is expected that the time between R&D and production will shrink from 3–5 years today to 2 years. (PwC 2019). The automotive industry has traditionally relied on physical prototyping (Pittman 2019). Now VR allows designers to visualize a whole vehicle without physical prototyping, or 'see' inside a new engine during its design. In manufacturing, Industry 4.0 technologies are having an impact from the level of the individual worker to the global supply chain (e.g. Airbus already uses cobots). More broadly, robots can raise quality, and automate material movements, repetitive tasks and predictive maintenance.[2]

Enhanced connectivity can greatly improve customer service. It is already critical in modern airliners (e.g. an A350 transmits 30 gigabytes of data daily). Aeroengine manufacturers are already delivering servitization to customers, monitoring every engine in flight in real time, detecting faults and predicting its maintenance requirements. It is projected that in aerospace advanced data management will allow the development of digital twins for every aircraft produced that trace every component of an aircraft from its manufacture to disposal and recycling (e.g. Economist 2019). Finally, although automobile customers and airlines already expect to receive vehicles built to their exact specifications, VR has the potential to enhance their experience further (e.g. enabling car buyers to 'test drive' the exact vehicle they intend to buy).

Of course, Industry 4.0 innovations create challenges as well as opportunities (Klaas 2019). They create competitive pressures, major technological uncertainties and pose an existential threat to firms that lag behind or make poor investment choices. Automotive and aerospace are engineering-based industries, which inevitably draw on many technologies. In a time of rapid technological change, there are leads and lags in technological development. One technology may apparently offer enormous immediate potential, but be hindered by the slow pace of other technologies upon which it is dependent (e.g. the time taken to develop additive manufacturing for composites). Technological complexity makes for difficult decision making. Not only are the various Industry 4.0 technologies advancing very quickly, but both industries need to incorporate a range of other technologies into their products. For example, there are currently three major technological drivers in automotive – electric vehicles, connected and autonomous vehicles and on-demand mobility services (KPMG 2019b). The impacts of Industry 4.0 technologies are also dependent on people's responses to them. Consumers may not accept some innovations based on Industry 4.0 technologies because they do not offer sufficient individual or societal benefits. Citizens may object to their broader social consequences. Reservations about an innovation may arise because it is seen as unsafe (e.g. consumers may be reluctant to trust autonomous cars, let alone driverless air taxis). Worse still, customers may be unwilling to use a product if it actually proves unsafe is use (e.g. the Boeing 737 MAX). Public attitudes towards these technologies are mediated through public policies that may help or hinder their development (e.g. with subsidies and preferential traffic regulations for electric cars). What may be technologically possible may not be socially and politically acceptable. Regulators are becoming more influential on automotive design (e.g. KPMG 2019a). Similarly, the 737 MAX disasters have led to a reassertion of regulatory authority by the FAA. Inevitably, such reactions are likely to have spatial impacts – particular products and services which are acceptable in some markets may be unacceptable, or be discouraged, in others (Financial Times Lex 2019). The complex relationships between these technological, managerial, social and political capabilities condition both the development and implementation of Industry 4.0 technologies in general and their spatial distribution. Hence regions have different technological and economic strengths and weaknesses regarding Industry 4.0 which are reflected: both in the diverse roles they play within the automotive and aerospace supply chains; and in enabling diversities of innovation as firms and research institutes collaborate across regional boundaries.

The impacts of Industry 4.0 innovations on established manufacturers

The reinforcement of established regions

Even when a technology is sufficiently developed, is economical, and consumers want it, for the individual firm there still remain critical decisions. The firm needs to consider – its own capabilities and resources, those within its supply chain and the impact of the technology on its competitive advantage. Many of the qualities that firms need to respond to the innovative demands of digitization are in common with the qualities they need to engage with other technologies. A firm will need not only technological knowledge and skills, but also a blend of managerial and commercial knowledge and skills to assess the potential competitive implications of the technology (e.g. Maisonneuve et al. 2013). Such decisions can be acute.

The issues facing a firm vary depending on whether it is an established player or a new entrant in the sector, and where it positions itself in the supply chain. In mature, high technology, engineering-based industries like these, established firms have enormous competitive advantages, in terms of their huge capital investments, their generations of R&D, their established supply chains, the experience and culture of their workforces, their brand recognition and so on. Established OEMs, and many of their Tier 1 suppliers, are global businesses, but they are also firmly rooted in particular regions by their human, physical and financial capital investments and so are unlikely to relocate core activities (e.g. Hickie 2006). Regions like Stuttgart in automotive or Seattle in aerospace demonstrate the competitive advantages to be drawn from their immense regional legacies.

It is important to recognize that customer companies and a local supply chain are not in themselves sufficient to create the fully functioning regional innovative ecosystem (Adner 2017). To flourish innovative companies need to draw on the services of governmental and non-governmental agencies. Ubiquitous in both industries is the need for research, design and manufacturing skills. Until the Covid outbreak the aerospace industry was experiencing both regional and national skilled labour shortages for postgraduate and graduate engineers, technicians, supervisors and factory workers (Azouzi 2014; Walker 2014; Washington State 2015). The growth of Industry 4.0 technologies creates an imperative in both sectors to build a workforce capable of innovating and implementing them. Both have recruited staff with digital skills from other sectors with more advanced digital skills (Aerospace Technology Institute 2018), but competition for talent is increasingly tight (especially regionally). Of course, these shortages are not likely to be felt equally across all types of employment or all regions. PwC (2019) forecast a halving of assembly line workforces in automotive due to automation and new types of shared mobility vehicle, whereas the number of software engineers may increase by up to 90%. Hence, plants focused on manufacturing may lose process workers, but suffer shortages of digital skills.

To help train and rebalance their workforces, firms are often heavily dependent on their regional education and training infrastructures. Clearly, OEMs and Design and Build suppliers generally will require a greater supply and variety of specialist skills, but even simple Build-to-Print suppliers need their own engineers with highly developed digital skills, if they are to fit themselves into highly integrated supply chains and the Internet of Things. Hence firms are reliant on educational and training institutions (e.g. schools, technical colleges, universities) to ensure they are competitive in Industry 4.0 technologies. Furthermore, in these highly innovative regions automotive and aerospace firms combine with their non-commercial partners in regional cluster organizations (e.g. HEGAN for Basque aerospace and automotive-bw for the Stuttgart auto industry). These organizations often play a key role in both coordinating and providing education and training, and in developing the industrial culture. They can lobby and negotiate with local education and training providers and with national and regional governments to ensure that the education system is providing for the industry's future demands for skilled labour (e.g. by ensuring school leavers have a good STEM education). In addition, cluster organizations play a role in marketing the industry in their region, both to help sell its products and to attract new OEMs and suppliers to set up in the region.

In essence, there is strong evidence that automotive and aerospace are still centred on long-established highly innovative regions with intense concentrations of innovative resources. Rather than moving, such businesses can choose to train and hire the talent and create the R&D facilities they need for the new technologies. They act as a magnet for startup companies that wish to supply Industry 4.0 products and services to their sector. Hence, there are regional agglomeration effects which both benefit established OEMs and their suppliers in a region, and attract new automotive or aerospace companies (Hickie, Jones, and Schloderer 2019). Over the past several decades, some globally significant new centres have developed (e.g. in Mexico in both automotive and aerospace), often based on FDI from major industry players. These can form the basis for regional ecosystems with their own local supply chains. Such supply chains may be made up both of indigenous supplier companies and by branches of global suppliers. Some of these regions (e.g. Singapore in aeroengines and MRO) (de Meyer et al. 2014) are developing as islands of innovation with a significant capacity to innovate that impacts beyond their own region (Hilpert 2019; Adner 2017). In general, these new innovative regional centres have not yet fully equalled, let alone displaced, the long established global centres as innovation hubs (except, arguably, Sao Joao dos Santos in Brazil for regional airliners). However, it is a clear intention of Chinese industrial policy (Ministry of Industry and Information Technology 2015) that China develops its own innovative centres in aerospace and automotive.

Drawing new regions into the supply chain

The 'gravitational' pull of established OEMs in well-established regions does not mean, though that, even within these global businesses, the geography of innovation is unchanging. Where a technology has a high degree of novelty is progressing quickly, and is critical to competitiveness, even globally dominant OEMs may need to look beyond the regions where they are long established. These global players can choose to locate new R&D investments in particular new technologies at new locations, which are established or developing global centres for those technologies. In Industry 4.0 technologies this has most obviously occurred in and around Silicon Valley. Major automotive and aerospace companies have been attracted by the existing innovative ecosystem there and invested in it in three key ways. Firstly, they have used their immense financial and technological strength to buy up startup ventures with innovative potential (e.g. in 2016 GM took over Cruise, a San Francisco-based autonomous vehicle startup, which is developing an autonomous electric Chevy Bolt). Secondly, they can invest heavily in an existing technology company, as Honda also did in Cruise in 2018. Or, thirdly, they can set up their own, focused R&D facility in a region they identify as having the necessary innovative ecosystem to enable their activities (e.g. Ford, Nissan and Airbus all have research facilities in Silicon Valley). Hence, the geography of innovation in both sectors has grown to incorporate Silicon Valley, whilst further strengthening Silicon Valley's regional competitiveness in Industry 4.0 technologies.

Simply being an established global player is no guarantee of technological success. Despite their accumulated knowledge and expertize, even well-established OEMs can lack sure footedness when introducing new products and technologies. Furthermore, being a legacy manufacturer can involve liabilities as well as strengths. Legacy status brings with it heavy past investment in existing technologies and the culture, values and careers built around them. Boeing's unfortunate decision to introduce the MCAS anti-stall system on its 737 MAX can be interpreted as a consequence of attempting to adapt a 1960s airframe. Legacy in the automotive industry can most obviously be seen in chronic oversupply amongst its most established players (e.g. Ford's sale of 6 of its 24 European plants; the sale of GM Europe to PSA). Here, consolidation and mergers are directly related to the need to share the cost of developing new vehicle platforms and speed their time to market using digital technologies. Closing outdated, inefficient plants inevitably reduce a company's footprint, both regionally and globally. Cultural legacies can lead to a certain conservatism and a reluctance to make the profound changes necessary to make full use of Industry 4.0 technologies. Such conservatism is a potential drag on established regions' ability to adapt to Industry 4.0, and hence on their long-term competitiveness. Currently, there is some evidence of conservatism in aerospace companies.[3] The evidence in automotive is more mixed. The new digital technology-based competitors in the automotive market have chosen to develop fully autonomous vehicles, regarding the more incremental R&D into semi-autonomous vehicles by some established automotive OEMs as legacy-based conservatism (Economist 2015). However, PwC has found that, globally, automotive was especially advanced providing 20% of the most digitally advanced manufacturers (digital champions) (PwC 2018).

Whilst legacy can be a source of strength for OEMs, global Tier 1 suppliers and technologically advanced suppliers, it is often a marked disadvantage for less advanced suppliers lower down the supply chain. Here, the adoption of digital technologies tends to be slower and less advanced. This can be a profound risk for smaller companies, which may lack the financial, technological and managerial resources to engage with radical technological developments (Hickie, Jones, and Schloderer 2019). Such firms often supply on a build-to-print basis (not owning critical IPR), are heavily invested in existing technologies, and may be manufacturing components which are highly commodified. Such firms can simply be unaware of the impending technological changes imminent in their supply chains. Their low margins mean they may lack the investment funds necessary to innovate, and cannot afford the specialist labour they would need.[4] It was clear some years ago in aerospace that significant numbers of these established SME suppliers were ill-prepared and struggling to keep pace with the key new technologies. This applied even to SMEs in regions where the industries were well established (e.g. Maisonneuve et al. 2013). Both sectors have global supply chains with Tier 3 or 4 suppliers in lower wage economies whose competitive advantage lies significantly in lower labour costs rather than in their own, or their region's, advanced technological skills (e.g. many suppliers of automotive parts in Eastern Europe). Industry 4.0 threatens to displace such suppliers with companies employing advanced technologies and potentially with even lower costs. The demise of these weaker suppliers could have a significant impact the geography of manufacturing, but would have much less of an impact on the geography of innovation, because the firms worst affected would not be strong innovators.

Industry 4.0 technologies and opportunities for new entrants and locations

New entrants

Despite the strengths of legacy OEMs and global suppliers there are, as suggested earlier, opportunities for new entrants in both sectors. More particularly, digitization offers opportunities for those developing whole vehicles incorporating digital technologies, and those developing as specialist suppliers of digital products and services. There are new OEMs designing and manufacturing complete new vehicles, which attempt a step change beyond what is currently being manufactured, perhaps even planned, by established OEMs. These firms, often startups, should not be seen simply as digitization specialists. In order to take on established players and produce highly innovative vehicles they will draw widely on the range of technological advances being employed in their chosen sector. Furthermore, these new OEMs are not necessarily attracted to set up in regions because of their previous expertize in automotive or aerospace. Tesla is a particularly interesting example. Although its competitive advantage currently rests primarily in the qualities of its electric engines and its marketing, it has chosen to locate in Silicon Valley and is developing its own vehicle connectivity to contribute to developing automated cars (Economist 2015). Its decision to locate there appears to have been driven by its proximity to digital expertize, which was reinforced by its purchase of the former NUMMI plant in Fremont. Similarly, the new Chinese electric car makers Byton and Nio are both using advanced digital technologies to enhance the appeal of their vehicles (advanced touch screen technology on the M-Byte and AI-enabled functions on the Nio). Yet the tasks facing new OEMs, whether Tesla or its Chinese competitors, are considerable. Tesla has only recently posted its first full year of profit, has ongoing problems with manufacturing quality, and has some way to go in developing its autonomous vehicle technology. Chinese car manufacturers, perhaps, have a more competitive opportunities, relying not only on new technologies, but also on long-term domestic market growth (28 million vehicles in 2018) and significant industrial policy support (not least in robotics) (McGee 2019). However, Chinese auto manufacturing startups still face a daunting prospect if they are to become globally competitive. For example, Nio recently undertook a new round of funding because of poor sales and cash flow issues.

In aerospace, producing large airliners is potentially technologically, financially and industrially even more taxing for new entrants. There have been two significant OEMs designing wholly original airliners to challenge Airbus and Boeing, COMAC in China and Bombardier in Canada. Both had to use composite materials and digital technologies to have even a chance of competing in global markets. In addition, both have relied heavily upon key suppliers and regions in the established aerospace supply chain. For example, while the COMAC C919 airframe is designed and built in China, many of its key systems are foreign designed. Although these systems may be assembled in China, and so help provide a knowledge base for future Chinese aerospace development, currently they embody innovations developed elsewhere. For example, its engines are from the Franco-US partnership CFM, while its flight controls are from the leading US supplier Honeywell. Despite being able to call on world leading suppliers, both COMAC and Bombardier have found the development of airliners ab initio very challenging. Both aircraft have experienced very lengthy delays in development, which in Bombardier's case led to the takeover of its airliner activities by Airbus as the project bled cash. Currently, the market leading capability to design large airliners remains with Boeing, Airbus and their supply chains.

A more promising market for new aerospace OEMs has been the cheaper and relatively simpler task of designing, and sometimes building, small aircraft relying heavily on digital technologies. These new aircraft are not intended directly to take on Boeing or Airbus in their key markets, but are regional executive jets (e.g. the Israeli Eviation Alice, Zunum Aero in Seattle), (Economist 2019), and air taxis – with or without pilots (e.g. e-Hang in China, Volocopter in Germany and Wisk in the US). Such companies are often supported by financial investments from established aerospace companies and/or digital entrepreneurs (e.g. Wisk has funding from both Boeing and Larry Page). In addition, major, digitally strong, delivery companies like Alibaba, Alphabet and Amazon are investing heavily in the development of pilotless delivery drones (e.g. Alibaba for remoter parts of China). What these developments have in common is that they rely heavily on digital technologies to allow autonomous or single pilot operation, as well as drawing on advances in composite materials and electric engines.

The locational choices of new entrants

New OEMs entering the automotive and aerospace industries, whose competitiveness rests on their Industry 4.0 technologies, have locational decisions to make about their manufacturing, design and R&D activities. Manufacturing decisions are influenced by a range of quite disparate factors, such as the supply of skilled industrial workers (especially those with sectoral experience), the proximity of suppliers, public policy and an element of serendipity (e.g. Tesla's decision to manufacture in Fremont). However, it is clear that when new automotive companies decide where to locate innovative R&D and design activities they are frequently attracted to well established regions, where there are well established innovatory ecosystems, networks and technologically skilled labour. So, for example, Nio's HQ is in Shanghai but it chose to set up: its design centre in Stuttgart; its software development centre in San Jose, California; and its motor racing centre in London. Similarly, Byton has design, engineering and software centres in Munich and Santa Clara, California – despite planning to manufacture in Nanjing. Similarly, Tesla's Fremont facility is close to Palo Alto, which can attract a ready supply of experienced and highly skilled software engineers more easily than in Detroit, even if at significantly higher wages. Tesla plans to build its European HQ in Brandenburg, close to Berlin, which again offers a ready supply of software engineers but, critically, has ready access to the Berlin's market potential for shared transportation. In aerospace the closest parallel is with OEMs designing small electric aircraft, often autonomous, for use in urban areas. Such companies are drawn to set up close to established R&D centres, to highly skilled engineering labour and often to major investors. For example, Volocopter is based at Bruchsal in Baden Wurtemburg, where it is close to Daimler-Benz, a major investor and a major research centre of the DLR (the German Government aerospace research body), as well as to a large pool of skilled labour from which it has drawn (e.g. its Chief Technical Officer came from Safran, a Tier1 aerospace supplier with a local base). Similarly, the Opener Blackfly electrical VTOL aircraft and the Kitty Hawk Cora flying car are both based in Silicon Valley, where their lead investor is based and expertize is abundant (Economist 2019).

Similarly, the introduction of new digital suppliers is adding a significant geographical diversification to innovation in both sectors. Industry 4.0 has its own established innovative geography, most obviously in Silicon Valley. Here established digital companies and new startups are developing innovative products that they can supply to both industries. For these new, innovative suppliers the most influential attractants, when choosing to locate their activities, are highly skilled labour, networks of suppliers, potential investors and potential customers. On the supply side, for example, Silicon Valley is an obvious location for highly specialist digital companies which want to supply AI-based driving platforms (e.g. Nauto and Plus.ai). AI-trained software engineers are essential to their competitiveness. However, some of the most highly specialized digital suppliers prioritize proximity to customers -OEMs and their established suppliers, – who provide a ready market for sector-specific Industry 4.0 innovations. For example, in Toulouse ID-Product, a robotics and automation specialist in cabin interiors, has identified a niche acting as a 'go-between' linking highly specialist SMEs lower down its supply chain and top tier manufacturers. It identifies SMEs capable of meeting very precise customer needs and, if necessary, assembling components from several suppliers for delivery to the OEM. Similarly, nearby Vodea supplies multimedia visualization of products for aircraft (e.g. for cockpits, for taxiing) employing AR, metadata management and machine learning. Stuttgart has a similar variety of new SMEs using Industry 4.0 technologies mainly to fill automotive market niches. For example, CVEP, created in 2016, provide electrical control systems with intelligent controls for vehicle electrics and electronics, whilst Drag&bot provides programming software for manufacturing robots. Around Detroit, Clinc is developing voice interface software for the automotive industry, and Advanced Collective Vehicle Solutions is developing communications technologies for vehicle entertainment, fleet tracking and security. Serendipity can also influence new firms' locational decisions (e.g. Detroit Flying Cars' founder has lived and worked in the region for decades).

The entry of new OEMs and digital suppliers into the automotive and aerospace sectors has the potential to make radical changes to the geography of manufacturing (e.g. the potential for major new automotive factories in China producing AI equipped electric vehicles). However, their impact on the geography of innovation is presently less marked. Even major new OEMs like COMAC, Byton and Nio in China are heavily reliant upon innovations developed in well-established innovative regions. Amongst new small OEMs and digital suppliers the concentration of innovative activity is, if anything, even more concentrated in established innovative regions – whether based on proximity to other digital businesses or to customers. Even where a new company appears to deviate from these tendencies, its departure can be less complete than it seems. For example, the Israeli aerospace OEM Eviation may appear to be a very long way from global aerospace innovation centres, but it has the support of a highly innovative indigenous digital sector. Furthermore, although its designs are Israeli, it test flies its aircraft in Washington State.

Potential changes in the geography of innovation in supply chains

Industry 4.0 technologies offer a range of opportunities for further offshoring in both sectors. For example, the exchange of vast amounts of production data has the potential to allow manufacturers to coordinate production processes in distant locations even more closely than before, and to avoid mismatches both in timing and in product specifications. This would allow experienced manufacturers to continue to export their manufacturing knowledge to lower wage economies. For example, if fully realized, the Internet of Things would allow robots at European or US car plants to communicate directly with robots at a supplier's plants in Mexico or Indonesia to coordinate production directly without human intervention. Indeed, automotive managers expect Europe's share of automotive manufacturing to continue to shrink further (KPMG

2019a). Similarly, an aerospace MRO facility in Dubai or Singapore could use 3D printed components using data transmitted direct from a manufacturer in North America to repair an aircraft. The Internet of Things could allow suppliers at distant locations to make process innovations and still ensure close coordination with their customers. However, it seems likely that the impact of these changes would be primarily on the geography of manufacturing. The locus of innovation would probably remain with OEMs and established Tier 1 suppliers, so the spatial impact of these developments would be primarily on manufacturing and routine services. Telemigration offers the possibility of offshoring technology-based services to low wage economies using instantaneous translation, enhanced connectivity, Big Data, AR and VR.

On closer inspection, however, even this seems an incomplete analysis. Advances in robotics require less of the semi- and low-skilled labour which gave developing economies much of their competitive advantage. This has the potential power to alter the economics of the supply chain. If labour costs become relatively less significant, in principle then, reshoring to newly automated plants in Japan, Western Europe or North America becomes a more attractive proposition, especially when they are located in regions with a strong transport and data transfer infrastructure. If components can be manufactured nearby by robots, and requiring only a small element of very highly skilled labour, it may make little sense to manufacture them thousands of miles away using the same or similar robots, especially as the Covid crisis has demonstrated the vulnerabilities of long supply chains. PwC has recommended that 'Manufacturers should be reviewing their supply chains and looking at where they can exert greater control, what they can bring closer to home and what technologies, such as 3D printing, can be used more to create greater agility in the supply chain' (PwC 2020b, 10; see also PwC 2020a). This will also enhance the industries' existing trend to operating with fewer suppliers. Furthermore, in a knowledge-based economy, competitive advantage lies primarily in people's heads, in teams and in organizations. In general, these are more likely to be found in well-established industrial regions with sophisticated innovative ecosystems. The nature of these innovations suggests that regions whose competitive advantage lies primarily in a ready supply of cheap unskilled, semi-skilled and technician labour may be more vulnerable to advances in digitalization. Hence, these innovations could benefit those well-established automotive and aerospace regions, with strong innovatory ecosystems in Europe, Japan or North America, provided that they are not overburdened with legacy investments and cultures.

Conclusions and discussion

It is clear that Industry 4.0 technologies are having quite profound effects upon product and process developments in both automotive and aerospace. For example, the switch away from direct sales to servitization in the aeroengine market is currently having profound adverse effect on Rolls Royce because so many airliners are not flying due to the Covid crisis. It is also clear that academics, industry observers and industry insiders fully expect that Industry 4.0 technologies will have even more profound effects on both sectors in the 2020s and 2030s (e.g. the introduction of semi-autonomous and autonomous vehicles, the introduction of MaaS in major conurbations). But it is also clear that we are currently still only in the foothills of such developments, still unable to forecast precisely the directions that innovation will take, nor to estimate with precision the technological or economic success, or otherwise, of particular innovations (e.g. autonomous cars or air taxis). This position is not unusual at this stage in the development of new technologies (e.g. it has parallels with the development of biotechnology in, say, 1990).

Furthermore, these new technologies are not simply sources of boundless innovative opportunity for firms. They also bring with them challenges and uncertainties. Advances in Industry 4.0 technologies are both potentially disruptive and expensive. Furthermore, they are not a single technology, but multiple related technologies, often developing rapidly. Hence, they involve risks, even existential risks, for firms. Levels of risk are enhanced because innovation in automotive and aerospace also involves other technologies. For example, both see innovations in engine technology as critical to their future development, as well as developments in vehicle autonomy. Finally, the economic success or failure of these innovations is not simply a technological or supply side matter. It is dependent on the societal response to them by consumers and regulators.

Despite these technological, economic and societal uncertainties, some of the ways in which Industry 4.0 is impacting on the geography of both sectors are becoming clear. In regions that are already technologically advanced firms are reinforcing their competitiveness as OEMs and leading suppliers by investing in digital technologies. In addition, these companies and their regional ecosystems act as powerful attractants to new, innovatory OEMs and to specialist digital suppliers. However, even among established players there is also an expansion of their innovatory footprint to encompass new regions which are already leaders in digital technologies (most clearly Silicon Valley), both by drawing world leading digital companies into their supply chains, and by establishing their own specialist digital research facilities in those regions. This new collaboration is complementary and enhances the competitiveness of the regions involved. The other key point of geographical diversification by established players in their desire to develop and test products that need early consumer and regulatory acceptance in major potential markets (e.g. autonomous vehicles in Berlin or Shanghai).

Digitally focused new entrants, whether OEMs or suppliers, usually choose between setting up in the strong existing aerospace and automotive regions, or locating in regions which are primarily strong in digital technologies, and so benefit from the strong local innovative ecosystem, local potential investors and a supply of highly skilled labour. There are, however, exceptions, most obviously in China, where national industrial policy and the prospect of a large home market require new entrants to set up domestically. Here, though, there is a difference between aerospace and automotive. Chinese aerospace companies design and build aircraft domestically, though they draw heavily and critically on overseas suppliers. Chinese automotive companies manufacture at home, but often conduct R&D and design in innovative regions abroad. However, both Chinese automotive and Chinese aerospace illustrate a key finding of this study. Currently, whilst manufacturing may be globally dispersed, leading edge R&D and design tends to be more focused on established regions with strong innovative ecosystems, whether based on aerospace, automotive or digital technologies. The future geography of manufacturing is potentially more uncertain. Whether supply chains continue to be located offshore, or are subject to reshoring, depends upon the continued progress of digital technologies (notably to support the Internet of Things), the digital and managerial capabilities of firms in dispersed supply chains, and OEMs' appetite for supply chain risk.

These findings are not only relevant to these two industries. They raise issues relevant to the analysis of the technological, economic and spatial impacts of Industry 4.0 technologies on other sectors especially, but not only, in manufacturing. In particular, they demonstrate: the need to understand Industry 4.0 technologies within the context of the other sector-based technologies with which they are applied; the impact of sectoral histories and company legacies upon their capacity to use Industry 4.0 technologies; the continuing prominence of well-established knowledge-based, regional ecosystems and their capacity to co-opt new technologies, whether locally or from elsewhere; the capacity for new centres of regional excellence to develop based on digital technologies; and that Industry 4.0 technologies are still at quite an early stage in their development so that much about them remains uncertain.

Notes

1. Predicting the collective future impact of a wide variety of interconnected technologies like Industry 4.0 is intrinsically uncertain. Whilst very broad directions of travel for both sectors are relatively clear, it is also evident at the time of writing (December, 2020), that the economic consequences of the Covid 19 virus pandemic have added further layers of uncertainty – most obviously for aerospace given the projected 3–5 year dip in passenger numbers.
2. For example, General Electric uses 3D printing to make weight saving shapes (e.g. voids) that are difficult to make using conventional manufacturing.
3. For example, in 2017 most UK aerospace companies were introducing digital technologies incrementally, rather than as a basis to transform their businesses, Aerospace Technology Institute 2018. See also Maisonneuve et al. 2013.
4. For example, a third of European automotive sector companies lack confidence in their ability to the necessary recruit talent (KPMG 2019a).

Disclosure statement

No potential conflict of interest was reported by the author(s).

References

Adner, R. 2017. "Ecosystem as Structure: An Actionable Construct for Strategy." *In: Journal of Management* 443 (1): 39–58. doi:10.1170/0149206316678451.

Aerospace Technology Institute. 2018. *Insight. The Economics of Aerospace: The Evolving Aerospace R&D Landscape, Cranfield, December.*

Alcacer, J., and W. Chung. 2007. "Location Strategies and Knowledge Spillovers." *Management Science* 53 (5): 760–776. doi:10.1287/mnsc.1060.0637

Azouzi, R. 2014. "Training the Next Generation." *Aerospace* 41 (10): 36–37.

Baldwin, R. 2016. *The Great Convergence: Information Technology and the New Globalization.* Cambridge, MA: Harvard University Press.

Baldwin, R. 2019. *The Globotics Upheaval. Globalization, Robotics and the Future of Work.* London: Weidenfeld and Nicolson.

Brettel, M., N. Friedrechsen, M. Keller, and M. Rosenberg. 2014. "How Virtualization, Decentralization and Network Building Change the Landscape: An Industry 4.0 Perspective in World Academy of Science, Engineering and Technology." *International Journal of Information and Communication Engineering* 8 (1): 37–44.

Ceruti, A., P. Marzocca, A. Liverain, and C. Bil. 2019. "Maintenance in Aeronautics in an Industry 4.0 Context: The Role of Augmented Realityand Additive Manufacturing." *Journal of Computational Design and Engineering* 6 (4, October 2019): 516–526. doi:10.1016/j.jcde.2019.02.001

de Meyer, A., P. Williamson, H. Joshi, and C. Dula. 2014. *Rolls-Royce in Singa- Pore: Harnessing the Power of the Ecosystem to Drive Growth.* Singapore Management University.

Economist. 2015. *Upsetting the Apple Car.* 19 February 2015.

Economist. 2018. *Autonomous Vehicle Technology Is Advancing Ever Faster.* 1 March, 2018.

Economist. 2019. *Technology Quarterly, The Future of Flight.* 1 June 2019.

Financial Times Lex. 2019. *Aviation/Climate Change: Plane Speaking.* 22/23 June, 2019, p. 12.

Grendel, H., R. Larek, F. Riedel, and J. Wagner. 2017. "Enabling Manual Assembly and Integration of Aerospace Structures for Industry 4.0 – Methods." *Procedia Manufacturing* 14 (issue??): 30–37. doi:10.1016/j.promfg.2017.11.004

Grybowska, K., and A. Lupicka. 2017. "Key Competences for Industry 4.0." *Economics and Management Innovations* 1 (1): 250–253. doi:10.26480/icemi.01.2017.250.253

Haskell, J., and S. Westlake. 2017. *Capitalism Without Capital: The Rise of the Intangible Economy.* Princeton, NJ: Princeton University Press.

Hickie, D. 1991. *Archipelago Europe – Islands of Innovation. The Case of the United Kingdom.* Brussels: Commission of the European Communities.

Hickie, D. 2006. "Knowledge and Competitiveness in the Aerospace Industry." *European Planning Studies* 14 (3): 697–716. doi:10.1080/09654310500500254

Hickie, D., N. Jones, and F. Schloderer. 2019. "Globalization, Competitiveness and the Supply of Highly Skilled Labour in Civil Aerospace." In *Diversities of Innovation,* edited by U. Hilpert, 275–295. Abingdon: Routledge.

Hilpert, D. 1991. *Archipelago Europe – Islands of Innovation. Synthesis Report.* Brussels: Commission Of the European Communities.

Hilpert, U. 2019. "About Socio-Economic Development, Technology and Government Policies Diversities of Innovation." In *Diversities of Innovation,* edited by U. Hilpert, 3–32. Abingdon: Routledge.

https://techcrunch.com/2016/09/14/cruise-has-around-30-self-driving-test-cars-on-roads-right-now/. 2016.

Klaas, B. 2019. *The European Automotive Sector Facing Unprecedented Change, The Engineer.* 15 October, 2019.

KPMG. 2018. *Autonomy Delivers: An Oncoming Revolution in the Movement of Goods.*

KPMG. 2019a. *Global Automotive Executive Survey* 2019.

KPMG. 2019b. *Mobility 2030: Transforming the Mobility Landscape.*

Maisonneuve, F., M. Santo, H. Menard, and Schmidt. 2013. *Internationalization and Competitiveness of Aerospace Suppliers, a Joint Analysis of Germany and France, h&z Unternehemensberatung AG, Munich and Kea and Partners.*

McGee, P. 2019. *Crisis-hit Pronto Vows to Stay on Track, Financial Times.* 3 September, 2019.

McKinsey Global Insitute. 2017. *A Future that Works: Automation, Employment and Productivity.* January 2017.

Ministry of Industry and Information Technology. 2015. Made in China 2025, Beijing, People's Republic of China.

Pittman, B. 2019. *Disruptive Technologies Are Transforming Automotive Design, The Engineer.* 10 December, 2019.

PwC. 2018. *Asia Pacific Manufacturing Companies Champion Digital Transformation; Gap with Americas and EMEA set to widen.* 4 October, 2018.

PwC. 2019. *Strategy&.. Shared Mobility and Automation will Reshape the Auto Industry by 2030.*

PwC. 2020a. *Strategy&. Covid-19 UK Industry Focus. Where Next for Automotive?*

PwC. 2020b. *Strategy&.. Covid-19 UK Industry Focus. Where Next for Aviation?*

Rotman, D. 2017. The Relentless Rise of Automation. *MIT Review.* 13 February, 2017.

Siedsschlag, I., D. Smith, C. Turcu, and X. Zhang. 2013. "What Determines the Location Choice of R&D Activities in Multinational Firms?" *Research Policy* 42 (8): 1420–1430. doi:10.1016/j. respol.2013.06.003

Theorin, A., D. Bengtsson, J. Provost, M. Lieder, C. Johnsson, T. Lundholm, and B. Lennartson. 2017. "An Event-Driven Manufacturing Information System Architecture for Industry 4.0." *International Production Research* 55 (5): 1297–1311. doi:10.1080/00207543.2016.1201604

The Times Tempus. 2020. *Aviation Industry not Going Anywhere.* 1 May 2020, p. 44.

Valachek, J., L. Bartalsky, O. Rovny, D. Sismisova, M. Morhac, and Loksik. 2017. The Digital Twin of an Industrial Production Line Within the Industry 4.0 Concept. In: *Proceedings of the 21st International Conference on Process Control (PC)*, Strbske Pleso, Slovakia, 6-9 June, 2017, pp. 258–262.

Walker, C. 2014. "Mind the Skills Gap." *Aerospace* 41 (4): 18–21.

Washington State. 2015. *Workforce Training and Education Coordinating Board. Aerospace Manufacturing Skills. Supply, Demand and Outcomes for Washington's Aerospace Training Programs.* Olympia: Washington State.

World Economic Forum. 2016. *The Future of Jobs, Employment, Skills and Workforce Strategy for the 4th Industrial Revolution*, Geneva. January, 2016.

Wright, R. 2019. *Next Wave of Automation Ushers in Co-worker Robots*, Financial Times. 30 May, 2019.

The transition of regional innovation systems to Industry 4.0: the case of Basque Country and Catalonia

Francesco D. Sandulli ⓘ, Elena M. Gimenez-Fernandez and Maria Isabel Rodriguez Ferradas

ABSTRACT

The work looks at how regions design policies to facilitate the transition of regional innovation systems to Industry 4.0. The research analyses how regional Industry 4.0 policies should take into account the integration of the position of the regional productive system into international supply chains, the games of legitimacy and power of the actors involved in the innovation system, the institutional structures that allow the exchange of knowledge on Industry 4.0. between the agents and the connection between the synthetic knowledge base (engineering driven) and the analytical knowledge base (science driven) of the region. Through a detailed case study of the background, structure and impact of Industry 4.0 in the Spanish regions of the Basque Country and Catalonia, the work demonstrates how it is not possible to define a policy of promoting Industry 4.0 that is generalizable to all regions and how each region will have to adapt the design and implementation of its Industry 4.0 policies to the specific characteristics of its regional innovation system. Therefore the replication of policies from other regions will not be an effective mechanism for promoting Industry 4.0 since the transition to Industry 4.0 is a very regional specific and diverse process.

Introduction

In situations where the prospect of a revolutionary change in the technological paradigm arises, national and regional governments cannot make optimal allocations of public resources to the various elements that could potentially favour or impede the adoption and exploitation of the new technological paradigm. Truthfully, governments cannot predict the intricate network of endogenous and interconnected initial, intermediate and final effects associated with the technological revolution on social and business structures, monetary flows or the development of other complementary or adjacent technologies (Bostrom 2007; Nightingale 2004).

This blind decision-making usually causes public bodies to try to play all lottery numbers. This behaviour can be seen in the case of innovation policies related to Industry 4.0 in the Spanish regions. In fact, 15 of the 17 Spanish regions established Industry 4.0 as one of their priorities in their Smart Specialization Strategy plans, regardless of the region's actual capabilities to achieve this goal.

The main objective of this work is to investigate the elements that may favour or prevent the transition of a regional innovation system to Industry 4.0. To achieve this goal, the work compares the adaptation processes to Industry 4.0 of the two most industrialized regions in Spain: Basque Country and Catalonia.

Industry 4.0 in Spain

In Spain, the first attempts to transition to an early conceptualization of Industry 4.0 took place at the beginning of the twenty-first century in the automotive sector. This industry accounted for 60% of all robots in Spain in 2008. However, these robots were mostly concentrated in the manufacturer's plants and not on suppliers' plants. Most of the robots used in these plants performed welding activities and had a low level of connectivity, so they could not be considered as CPS. The first implementations of CPS in Spain started in 2010 when firms increased the use of manipulative robots in front of welders and Telefonica, the main telecommunications operator in Spain, launched the first M2M (Machine to Machine) communication services. At this early stage, there was also a first core of pioneering companies in Spain that offered infrastructure or services linked to CPS technologies, mainly oriented towards business models based on the Internet of Things (IoT). However, at that time, the demand for Industry 4.0. technologies was still very weak and many of these pioneering companies went bankrupt or reoriented their business models.

In terms of public policies, whereas during the 2010s some countries already launched Industry 4.0 plans such as Industry 4.0 in Germany, Advanced Manufacturing Partnership in the US or Catapult High Value Manufacturing in the United Kingdom, the Spanish policies for the development of the information society (Plan Avanza between 2006–2010 and Plan Avanza 2 between 2011-2015) focused more on the adoption of less complex information technologies such as the web or ecommerce. In 2015, the Spanish government launched the Connected Industry Plan to promote the digital transformation of the Spanish industry. The plan was structured around four lines of action: Awareness and Training, Collaboration Platforms for the Promotion of Industry 4.0, Creation of a critical mass of Spanish companies that provide technology and services linked to Industry 4.0, Creation of conditions favourable to the realization of Industry 4.0 projects including regulation, financing and advice. Due to the high decentralization of the Spanish State, each of the 17 Spanish regions has developed its own plan to implement each of these lines of action.

Today, most of the Spanish firms adopting Industry 4.0 are suppliers integrated in international supply chains, mostly in the automotive[1] and aerospace[2] industries. As typically happens in these industries (Rodríguez Ferradas, Alfaro Tanco, and Sandulli 2017), the innovative impulse of global OEMs is driving the adoption of new additive manufacturing or 3D printing technologies by Spanish suppliers, especially in Tier 1.

Basque Country and Catalonia are the most advanced Spanish regions in terms of adoption of Industry 4.0. In the rest of Spain the adoption of Industry 4.0. technologies is in its early stages and with very heterogenous regional characteristics. For instance, in the region of Castilla-León there is some development in the largest companies in the region such as Renault (automotive), Campofrío (AgriFood) and in local suppliers of automotive companies or machine tool companies in the Basque Country. The Madrid region is characterized by industry 4.0 projects in a few suppliers in the aerospace sector, but above all by concentrating the largest number of companies providing Industry 4.0. IT solutions (Software and Consulting Services linked to Big Data, Simulation or Artificial Intelligence). In the Valencian Community, apart from local suppliers of Ford´s plant near Valencia, the adoption of Industry 4.0 is led by vertical research centres specialized by industry such as Furniture, Toys, Plastic, Ceramics, Food, Textile or Footwear (the industries grouped in clusters in the region) within REDIT (Network of Technological Institutes of the Valencian Community), following the well-established tradition of outsourcing R&D by Valencian SMEs to research centres (Albors-Garrigós, Rincon-Diaz, and Igartua-Lopez 2014). In Andalusia the adoption of Industry 4.0 initiatives is linked to companies in the supply chain in the aerospace sector and to a lesser extent in the naval sector, since the region has Airbus plants and the largest shipyard in Spain, belonging to the public company Navantia. In the rest of Spain, with the exception of some rare initiatives linked to automotive or aerospace supply chains in Asturias, Galicia or Aragon, the adoption of technologies linked to Industry 4.0 is still very scarce.

Industry 4.0 in the Basque Country – the role of existing structures in industry and institutions

The Basque Country is together with Catalonia the region with a higher level of industrialization in Spain. The advanced manufacturing strategy is a priority area of the region´s Smart Specialization Strategy as reflected in the Basque Industry 4.0 Plan. The implementation of this plan is led by SPRI, the Development and Infrastructure Department of the Basque Government. The plan has three action pillars: Basque Digital Innovation Hub (a network of advanced manufacturing assets and services, including also infrastructure for training, research, testing and validation available to Basque companies), Training for Employment and New Business Models.

The specificities of the current adoption trend of Industry 4.0 in the Basque Country is the result of several years of industrial restructuring policies which started in the 1980s. For instance, Industry 4.0 is predominantly adopted in this region by the machine-tool and metallic products industries. This sectoral specialization of Industry 4.0 is explained by the significant integration in international supply chains of these two industries, which was promoted by the first regional policies of industrial transformation during the 1980s.

A second characteristic of the Industry 4.0 in the Basque Country is that the cooperation and knowledge exchange between industries is not as strong as expected. Since the 1990s, the regional industrial policies were cluster-based, making the Basque Country one of the world pioneers in the establishment of a Porterian cluster policy (Ketels 2004; Aranguren et al. 2014; Konstantynova 2017). These policies allowed the consolidation of some highly competitive specialized agglomerations. However, this high level of specialization created significant challenges to the adoption of Industry 4.0. In fact, Industry 4.0. projects typically blur the boundaries between industries and require cooperation across different sectors. However, the past cluster policies created islands of innovation where firms had little tradition of cooperation with firms in other clusters. Therefore, the successful implementation of Industry 4.0. now requires to establish almost from scratch horizontal cooperative relationships with firms across different industries. Similarly, the implementation of these projects needs the coordinated action of Basque clusters, which were organizations that in the past were used to share some knowledge with other clusters, but were not so familiar with the implementation of inter-cluster projects.

While in the 1980s, industrial policy in the Basque country adopted a traditional top-down approach, after the 1990s the Basque government started to cooperatively design and implement industrial policies (Ahedo 2004; Aragón et al. 2014). This public-private cooperation is clearly a pillar of the Basque Industry 4.0 plan, as shown by the Basque Digital Innovation Hub, which is a network of private-public collaborations composed of infrastructures, laboratories, equipment, software, and innovative and scientific-technological capabilities in the advanced manufacturing environment created with the aim of providing industrial companies, especially SMEs, with the technological capabilities necessaries for Industry 4.0.

Despite the strong public-private ties, cooperative arrangements between private actors in the Industry 4.0 field are not so common. Again, this reflects the fact that in the Basque Country there is little tradition of interfirm co-operation (with the very important exception of the Mondragon Corporation) (Moso and Olazaran 2002; Morgan 2016).

In order to deal with this problem, the Basque Government launched the Innobasque agency in 2007, as a region-wide framework through which to stimulate cooperation networks for innovation. Nevertheless, these policies have been more successful in the promotion of public-private rather than purely private projects as shown by the results of the programme Basque Industry 4.0.[3]

A successful transition of regional innovation systems to Industry 4.0 requires a solid base of frontier scientific knowledge in the field of IT solutions, robotics and automation (Götz and Jankowska 2017; Castelo-Branco, Cruz-Jesus, and Oliveira 2019). The Basque Country is currently struggling with the absence of this knowledge base. In fact, the region has been dragging the problem of little indigenous supply-side capacity in the science-based sectors for years (Navarro et al. 2012). While, universities and startups should normally assume a leading role in creating this frontier knowledge base in Industry 4.0, the Basque Country cannot boast neither strong universities (Morgan 2016) nor a strong base of innovative startups related to Industry 4.0 technologies (Orkestra 2018).

The region is also being challenged by the problem of spreading the local Industry 4.0. knowledge across the whole industrial system, since research centres integrated into the Basque Excellence Research Centres or Cooperative Research Centers tend to cooperate in Industry 4.0 projects more with large corporations and not with SMEs. In this sense, the fact that SPRI leads the development of Industry 4.0 in the Basque Country has biased its policy towards industrial development and not so much towards the integration between the science and technology system and the business structure, an area in which Innobasque should have had more prominence in the last decade (Morisson and Doussineau 2019). This limited attention to the connection between the science and technology system and industrial firms is a recurrent problem in Basque industrial policies over the years (Borrás and Jordana 2016), despite the efforts of programmes such as Basque Industry 4.0 mentioned above.

The region not only suffers from mismatches between the science and technology system and the industrial structure, but also lacks a critical mass of startups focusing on frontier technologies linked to Industry 4.0 with some exceptions in industrial cyber-security. To solve this problem, in 2016 the Basque Government launched Bind 4.0, a public-private startup accelerator designed to build relationships between large industrial companies and Industry 4.0 startups. The programme consists of subsidizing with a maximum of 150,000 euros implementation projects Industry 4.0 technologies in a large Basque industrial corporation.

One of the goals of the programme was to attract to the Basque country startups from other locations in the world. However, during the first two years of the programme, most of the startups of the programme were Basque. Only after the third year, the project started to attract a larger proportion of non-Basque startups. One of the factors that is making difficult the implementation of non-Basque startups in the region is the limited duration and scope of the projects supported by Bind 4.0. which do not justify the investments needed to move the firm to the Basque Country. Another weakness of the region in terms of lack of critical mass of startups is the small number of Basque industrial companies that have created spin-offs aimed at exploring new business models based on hybridization between the industrial knowledge of the parent company and the technological flexibility of the spinoffs.

After a detailed review of Bind 4.0. projects, we have observed that they are usually non-strategic projects for the large companies participating in the programme. The non-strategic nature of these projects create little internal traction towards the trans-formation to Industry 4.0 of these large industrial companies. In this sense, most of the projects analysed focus on improvements, often marginal, of the production process and not so much on the creation of a portfolio of new products or business models supported by Industry 4.0. technologies. This is not a problem of Bind 4.0, but in general terms most Basque companies adopting Industry 4.0 technologies are more focused on process innovation and incremental product innovation rather than on radical product or business model innovation.

This behaviour actually reflects the fact that the relative strength of the Basque Country rests primarily on process innovation and its position is weaker in product inno-vation, a type of innovation more closely linked to greater autonomy and decision-making power at firms (Orkestra 2019). In fact, a relatively limited number of manufac-turing firms have their own product, while most produce components, tools, materials or machines to be used in the manufacture of other final products. That means that archi-tectural product innovation in the value chains takes place elsewhere, typically large cor-porations in the automotive or aeronautics industries (Hickie 2020), and therefore managers of Basque firms have little freedom to use Industry 4.0. to introduce significant innovations into products with more value added. This dependence on global producers that control global supply chains represents an important challenge (Hickie, 2020) in the Industry 4.0 strategy of the Basque country.

A skilled workforce is one of the pillars of adaptation of regional innovation systems to Industry 4.0. The Basque Country is characterized by being the Spanish region that has paid the most attention to vocational training in the past. The regional government plans that an important proportion of the skills necessary for Industry 4.0. will be developed through vocational training. The Basque Fifth Vocational Training Plan focuses 'on the fourth industrial revolution, intelligent systems, advanced technologies and the new training and qualification needs of people for these complex environments'.

However, this plan does not solve some of the challenges related to Industry 4.0. skill development. First, it focuses on the skills of future workers, but pays less attention to adapting the skills of the current workforce. Second, Basque SMEs have little joint projects with vocational training centres that could be used to transform the skills of their workforce. This is not a new problem, since Albizu et al. (2011) already shown that less than 10% of Basque manufacturing SMEs carry out joint innovation projects with vocational training centres. These SMEs have not developed specific plans to develop the Industry 4.0 skills of their workforce. North et al. (2019) confirms that these firms follow a rather reactive approach to digital transformation since these companies' employees are rarely empowered to experiment with digital transformation initiatives and there is a limited cooperation with outside partners. This weak lifelong training programmes imply the risk that, if the Industry 4.0 transforms the Basque industry, a significant percentage of skilled and unskilled blue workers would be difficult to redeploy to new technologies. To worsen this risk, the rigid social structure erected during the era of mass production in the Basque Country is still prevailing (Rodríguez-Pose 1999), which causes younger and better prepared workers face rigid labour markets which prevents social mobility.

While Industry 4.0. is a core pillar of Basque vocational training plans, the region lacks a clear strategic plan to a deep adaptation of most tertiary education to this new paradigm. This secondary role is explained by both the weak ties between the science and technology system and the Basque firms discussed above and the rigidity of the higher education system, compared to the higher capacity of the Basque government to quickly adapt the contents of vocational training to Industry 4.0.

In short, the transition to Industry 4.0 of the Basque Innovation System is heavily influenced by past policies and institutions which combine strengths such as public-private cooperative environment, the commitment to vocational training or the integration in global supply chains, with challenges to overcome such as the lack of a scientific frontier knowledge base, the lack of a critical mass of startups or the cluster structure that although favoured industrial specialization in the past, creates rigidities in a transversal transformational process such as Industry 4.0 Table 1.

Table 1. Main policy instruments to promote Industry 4.0 in the Basque country.

Instrument	Description
Basque Digital Innovation Hub	Connected network of advanced manufacturing assets and services Infrastructure for training, research, testing and validation available for companies
Assets 4.0	Digitally-linked network of R&D infrastructures, pilot plants and specialized know-how in different areas of advanced manufacturing
4.0 Subsidies	• **Hazitek**: Subsidy to carry out Industrial Research or Experimental Development projects • **Industrial Cibersecurity**: projects that address the convergence and integration of protection systems against cyber attacks for IT/OT (Information Technology / Operational Technology) environments in industrial manufacturing companies • **Basque Industry 4.0**: Subsidy for Industrial Research and Experimental Development projects that involve technology transfer from technology providers to industrial companies • **Ekintzaile**: Subsidy for maturing a business idea at a Business and Innovation Centre • **Gauzatu**: Subsidies for the creation and development of technology-based and/or innovative SMEs.
Bind 4.0	Acceleration programme connecting startups with large manufacturing, energy, healthcare and food tech companies and contracts worth up to €150,000

Industry 4.0 in Catalonia – bipolar innovation policies vs. skill deficits

Catalonia's economy is based on a long-standing industrial tradition, which has experienced a progressive transition to a new economic model. The innovative Catalan system is especially concentrated in the metropolitan area of Barcelona, where there is a dense and innovative industrial community of SMEs and an active presence of large multinationals, particularly in the biomedical, agro-food, chemical, automobile and ICT sectors.

In recent times, some sectors have grown in importance such as biomedical, agro-food or ICT, while some others such as automotive or textile face competitiveness challenges. This structural transformation poses a first challenge in terms of Industry 4.0 adoption, since some of the sectors which are adapting more quickly to this new paradigm, are declining in Catalonia.

The level of awareness of the Catalan industry about Industry 4.0. is still very low. For this reason, Catalonian innovation policies have a strong focus on raising awareness on the potentialities and implications of Industry 4.0.

One element that will be key in the discussion of the Catalonian Industry 4.0. policy is the fragmentation of the innovation system. The Catalan system of innovation is characterized by the presence of many heterogeneous agents connected through weak links (Riba Vilanova and Leydesdorff 2001). In the last decade, there have been several attempts to develop policies to achieve greater integration, but they have been fruitless (Borrás and Jordana 2016). For instance, at the beginning of the 2000s, the regional government promoted the creation of a large number of technology centres to facilitate technology transfer to firms. However, these centres were too many, too small and had low performance in terms of technological advancements.[4] The fragmentation of the system implies a high density of agents in the innovation system, usually small in size such as research groups, local technology centres or SMEs, and connected with a small number of local partners. In addition, these interactions in the Catalan innovation system are unstable over time as they are more short-term oriented.

In principle the leadership of the design of the policy of promoting Industry 4.0 in Catalonia rests with Acció, the public agency for the competitiveness of the Catalan enterprise, attached to the Ministry of Enterprise and Knowledge of the Catalonian Government. However, there are many regional and local public actors such as IoT Catalan Alliance, STEMCat, Open Industry 4.0, BCN Industry Hub among many others developing Industry 4.0 initiatives which are not always coordinated with Acció. For instance, Catalonia has at least 21 major organizations or networks at supporting the digital transformation of local industries and SMEs, which are not connected and fail to generate enough synergies between them (Rissola and Sörvik 2018).

The fragmentation of the innovation system reflects a weak industrial civil society and low levels of public-private collaboration in the design and implementation of industrial policies that causes a quite centralized decision-making (Ahedo 2018). The design of Industry 4.0 policies has been no exception to this top-down and centralized process, with little participation of the industrial and civil society in the assessment and design of Industry 4.0. policies in Catalonia.

Since more active policies to promote cooperative projects would be costly due to the fragmentation of the innovation system, Catalan policies follow a passive approach in creating cooperation networks, relying more on instruments to facilitate greater visibility of innovative agents in the system, such as the Marketplace 4.0, the Barcelona & Catalonia Startup Hub network, or the visibility of research centres and groups through the Tecnio network and seal.

From the early years of our century, the regional innovation policies created what is called as the bipolar structure of the Catalan regional innovation system (Bacaria, Borrás, and Fernández-Ribas 2004). On one hand, the Catalan government decided to bet decisively on the production of high-quality basic research produced by separate public research organizations under the umbrella of the Catalan Institution for Research and Advanced Studies (ICREA). On the other hand, innovation policies paid much less attention to firms' innovative activities, to the firms' external contracting of R&D services, and to the firms' interactions with public research organizations (Borrás and Jordana 2016).

This bipolar structure is influencing the regional transition to Industry 4.0. Despite the ICREA research projects have produced some amount of valuable basic research related to Industry 4.0 applications, Catalonia is struggling to transfer this knowledge to the productive system. On one hand, ICREA professors are more focused on outstanding research, but not as much involved in teaching, and therefore they contribute little to the development of an Industry 4.0. skilled workforce. On the other hand, ICREA Industry 4.0. projects typically have low technology readiness levels and therefore can hardly be integrated into the innovation funnels of Catalonian firms. Because of this incomplete knowledge transfer, Catalonia is not fully exploiting its potential capacity to compete on Industry 4.0. technologies. For example, according to data from the European Patent Office (Meniere, Rudyk, and Valdés 2017), Catalonia is not only lagging behind the European regions with the highest generation of Industry 4.0. patents, but also behind the Madrid region, even if Madrid has less capacity to generate basic knowledge in its public science and technology system.

The current RIS3CAT 2015–2020 Action Plan of the Catalonian government is not giving a prominent role to skills development or transformation to Industry 4.0. Within the activities of Acció, the training of SME managers stands out but through a national training programme promoted by the government of Spain. The current system of vocational training in Catalonia seems unprepared to promote the development of skills necessary for Industry 4.0 due to poor coordination between the different stakeholders, the lack of strategic planning, the absence of clear methods for the development of innovations through joint projects between SMEs and schools, and the fact that needs and competences demanded by SMEs are neither well identified nor integrated into the students' curricula (Hernández-Lara, Moral-Martín, and Brunet-Icart 2019).

Multinationals located in Catalonia play an important role in the transition to Industry 4.0. Specifically, we should highlight the central role of Hewlett Packard's global 3D printing innovation centre in the development of a 3D printing hub with more than 100 Catalan companies. Thanks to this initiative, Catalonia can be considered as a global leading provider of innovation linked to 3D Printing. In addition, this specialization in 3D printing technology fits the digital transformation plans of high growth industries in the region, such as chemistry, biotechnology or agri-food with the greatest growth potential in Catalonia such as. Finally, this specialization in 3D means that in Catalonia, compared to the Basque Country, there is a greater proportion of companies that rely on Industry 4.0. technologies to support product or business model innovation. Beyond the

3D printing hub, many of the Catalan industrial companies share with their Basque counterparts the small average size and their integration with little market power into international supply chains, where they have little capacity to capture the value introduced by product innovations supported by Industry 4.0 technologies.

Overall, Catalonia's transition to Industry 4.0 faces serious challenges associated with the fragmentation of the regional innovation system and the bipolarity of innovation policies. The region has the advantage of the tractor action of multinationals, but still the level of awareness over Industry 4.0 is low in the entire business sector. Finally, Industry 4.0 plans should pay more attention to skills generation and transformation Table 2.

Discussion – Divergent innovative opportunities and convergencies in skill deficits

The results of the analysis show significant differences between the transition of the regional innovation systems of Catalonia and Basque Country to Industry 4.0. The research reveals that there may be no canonical process or best practice to adapt a region to the technologies and business models associated with the concept of Industry 4.0., but possibly there is a diversity of different transition paths to Industry 4.0 depending on the specificities of each region. This first important result of the study supports the recent theory of Diversities of Innovation (Hilpert 2020).

A central contribution of the paper is to explain how the transition to Industry 4.0 depends on the pre-existing characteristics of the regional innovation system. Regions should be aware of how past industrial policies influence their Industry 4.0 transition. This result confirms the existing literature on the importance of path-dependence in the processes of change of regional innovation systems (Coenen, Moodysson, and Martin 2015). However, our analysis shows two different mechanisms of path-dependence. On the one hand, regional transition to Industry 4.0 depends on hard elements of the regional innovation system, such as the knowledge and skills base of the region and the size and sectoral structure of the local firms. On the other hand, transitioning to Industry 4.0. also depends on soft elements of the regional innovation system such as innovation policies and the nature of the interactions between actors in the system.

Table 2. Main policy instruments to promote Industry 4.0 in Catalonia.

Instrument	Description
Catalan Digital Innovation Hub	Network of local digital innovation hubs: IAM 3D Hub, Catalonia AI DIH, 5GBarcelona, Ecosystem Hub for high performance computing
Networking Platforms	CatLabs, IoT Catalonia Alliance, Catalonia Smart Drone, Robótica Industrial, Industrial Ring, Marketplace Industry 4.0
Directories of providers of Industry 4.0 capabilities	Tecnio, Barcelona & Catalonia Startups
Subsidies	**ICF Industry 4.0**: Soft loans to Industry 4.0 projects – SMEs (> 200,000 €) – large corporations (> de 800,000 €). **Industry 4.0 coupons**: Up to 14,000 € subsidies to SMEs for hiring services of Diagnosis 4.0 or implementation of projects 4.0 **TecnioSpring**: Subsidies to hire R&D workers for 2 years **Program INNOTEC (SMEs)**: Subsidies to R&D projects
Industry 4.0 Immersion and Industry 4.0 Course	Training programmes on general Industry 4.0 concepts for SMEs managers and workers.

In the specific case of the innovation system elements that regulate the transition to Industry 4.0, we have shown the role of the degree of integration of local firms into international value chains. Both in Catalonia and the Basque Country, the initial impulse to Industry 4.0 starts within these global supply chains. For instance, the more advanced adoption of Industry 4.0 in the Basque Country compared to Catalonia is explained by the fact that Basque firms are more integrated into global supply chains of industries which are early adopters of Industry 4.0. In Catalonia, some industries which are followers in terms of adoption of Industry 4.0, such as chemical, agrifood or textile, have a greater prominence in the regional economy. For this reason, Industry 4.0 policies in Catalonia need to compensate this weaker impulse of global supply chains towards Industry 4.0. with policies designed to raise awareness on the potential benefits of Industry 4.0.

The relative position of local firms in global supply chains will also influence of Industry 4.0 on process vs product and business model innovation. In both regions, a large proportion of the firms adopting Industry 4.0 are specialized component manufacturers. These firms are more likely to use Industry 4.0 technologies to foster process innovation and not so much product innovation and therefore will capture less value from Industry 4.0 innovations, compared to companies that have a greater influence on final product innovation.

Our research confirms tensions between regional innovation policy and pre-existing business models (Coenen, Moodysson, and Martin 2015). In these two regions, where companies have a weak position in global supply chains, Industry 4.0. policies are more focused on supporting pre-existing business models, instead of promoting business model innovation. In our research, this was especially evident in the case of the Basque Country, while in Catalonia firms belonging to the 3D Printing Hub relied more on business model innovation.

As in previous cases of long waves of technological change, regional governments have had to position themselves against the new wave of Industry 4.0 technologies to design instruments and provide resources to enable companies in the region to leverage these technologies to strengthen their competitiveness (Scherrer 2020). Our results show that this positioning is not an exogenous political process, but the result of the more or less intense interaction between the agents of the innovation system. For this reason, an important contribution of our research is that we need to study the transition to Industry 4.0 not as just one more combination of innovation policies or plans, but as a more general framework of action that involves the interconnected action of a multitude of agents. The ultimate objective of this framework of action, which Reischauer (2018) defines as discourse, is the institutionalization of the socio-technical changes associated with new technologies linked to Industry 4.0.

Our study describes two different approaches in the design of Industry 4.0. regional innovation policies. The top-down approach of Catalonia shows that the regional government holds the legitimacy over decisions on the objectives, actions and resources of Industry 4.0. policies. The bottom-up approach of the Basque Country shows how the legitimacy lies with public and private stakeholders affected by the definition of the objectives, actions and resources of the Industry 4.0. policies. However, analysing Industry 4.0. regional policies from a perspective of discourse and legitimacy allows to overcome the traditional dichotomy between top-down and bottom-up policies

(Fromhold-Eisebith and Eisebith 2005), insofar as the construction of the discourse linked to the adaptation of the regional economy to Industry 4.0 responds to an endogenous and continuous process of interaction between the different stakeholders and governments mediated by legitimacy games.

These games have recently been raised as the basis for the development of innovation policies (Flink and Kaldewey 2018). The frequency of interaction and above all the evolution of the distribution of power and legitimacy during the process of building the discourse between stakeholders and regional governments will allow Industry 4.0 policies to be placed in a continuum in which the top-down and bottom-up approaches are placed at the extremes. The mechanisms of legitimacy and results of this interaction between government and stakeholders depend on preconditions and characteristics of the different governments and stakeholders involved in the discourse. In the Industry 4.0. plans of the Basque Country, past policies and industrial history have given both large and small enterprises as well as vocational education greater legitimacy compared to higher education and the agents of the science and technology system. In Catalonia, research and technological centres and multinational corporations have greater legitimacy in innovation policies compared to SMEs or all educational institutions.

This different legitimacy distribution have strongly influenced the priorities and goals of Industry 4.0 policies in each region. Policies for the development of Industry 4.0 skills are a clear example of this phenomenon. The long tradition of vocational training in the Basque Country is reflected in the predominant role of this type of education in Industry 4.0 plans. However, both the less legitimacy of both the higher education system and the departments of the regional governments responsible for the development of the education explains why in both regions there is no systematic approach to a wide adaptation of university programmes to the future skills needed for Industry 4.0. In both regions, the transition to Industry 4.0 faces the challenge of weak links between the higher education system and the business environment. This weakness does not affect to joint research projects, but it put at risk the creation of a large base of Industry 4.0. skilled workforce.

Both regions share a low number of initiatives aimed at adapting existing workers to the new skills requirements of Industry 4.0. In part, this is explained by the fact that Industry 4.0 adoption in these regions is more related to process innovation or incremental product innovation, which compared to radical product or business model innovation typically requires a less ambitious adaptation of the workforce skills. The lack of lifelong learning plans in these regions poses a serious risk in the event of a strong industrial transformation driven by Industry 4.0, since these workers would not count on updated skills as a safety net.

Following the literature of complex systems (Feldman, Francis, and Bercovitz 2005), the regional transition to Industry 4.0 can be understood as an unpredictable and adaptive self-organizing behaviour of the players in the regional innovation system. One of the core elements of these adaptive interactions will be played by the institutional structures which favours effective knowledge exchange between different actors (Autio et al. 2018). In the Basque Country, the government employs more encompassing mechanisms to foster cooperation between agents, such as the Bind 4.0 programme, while the higher fragmentation of the Catalan innovation system requires relying on instruments to facilitate greater visibility of innovative agents in the system.

Finally, it would be interesting to study the regional transitions to Industry 4.0 taking into account how regional policies produce the necessary balance between synthetic knowledge base (engineering-based knowledge) and analytical knowledge base (science based knowledge) (Asheim and Coenen 2005). The evidence we provided shows that in the case of the Basque Country, science based knowledge in Industry 4.0 remains weak despite the efforts of the Basque government, while in the case of Catalonia the government has managed to create a very robust basic research system, but has not been able to create effective mechanisms to transform it into engineering based knowledge Table 3.

Conclusion – the diversities of regional opportunities of industry 4.0 in global contexts

This work has analysed the transition of the regional innovation systems of Catalonia and the Basque Country to the new paradigm of Industry 4.0. The work contributes to the

Table 3. Summarize of findings: Comparison of Industry 4.0 elements of the regional innovation system in the Basque Country and Catalonia.

Element of the regional innovation system in industry 4.0	Catalonia	Basque country
Industries most advanced on Industry 4.0 implementation	Automotive, 3D Printing.	Machine-tool, Metallic Manufacture, Automotive, Aerospace.
Type of Industry 4.0 driven innovation	Mainly process innovation because of integration on international supply chains and heavily dependent on OEMs plans to adapt to Industry 4.0. Some level of product innovation in the field of 3D printing and Additive Manufacturing	Mostly process innovation because of integration on international supply chains and heavily dependent on OEMs plans to adapt to Industry 4.0
Cooperation between agents in the implementation of Industry 4.0	Severe fragmentation of the innovation system. Weak public-private cooperation and loose and unstable networks.	Relevant public-private cooperation levels. Weak inter-cluster and interfirm cooperation.
Stage of implementation of Industry 4.0	Initial. Focus on creating awareness among industrial SMEs	Implementation. Focus on creating Industry 4.0 capabilities and transforming the industrial ecosystem
Governance of the Industry 4.0 policy design	Top-down approach to Industry 4.0 policy design with little participation of the stakeholders	Public-private cooperative process of design and implementation of the policies
Specificity of Industry 4.0 policy instruments	Mostly pre-existing instruments to foster innovation in general, not Industry 4.0 in particular	Mostly new and specific instruments created ad-hoc for the Industry 4.0 plan
Science driven support to Industry 4.0	Strong public basic research centres, but with little connections to the industry	Weak basic research capabilities and weak ties between universities and SMEs
Industry 4.0 Startup Ecosystem with Frontier Knowledge	Reasonable number of Industry 4.0 startups suffering from weak internal demand	Weak ecosystem of Industry 4.0 startups
Multinational Corporations Traction	Strong traction of multinational corporations in 3D Printing. Weak traction of multinational corporations through international supply chains.	Weak traction of multinational corporations through international supply chains.
Industry 4.0 Skill Development Strategy	No significant Industry 4.0 skills development strategy	Skills development strategy focusing on vocational training and future works. Less clear strategy related to lifelong learning and higher education.

literature on the processes of change in regional innovation systems showing how the hard and soft elements that define these innovation systems influence the process of adoption of Industry 4.0. The work emphasizes that the position of local companies within global supply chains should be taken into account as it will define the intensity and type of adaptation of local companies to Industry 4.0. In the specific cases of the Basque Country and Catalonia this position in supply chains influences the fact that companies in the region develop strategies of Industry 4.0 more focused on process innovation and not so much on product or business model innovation. Second, the work shows how one of the problems of the regions analysed is the absence of a clear strategy of skills development appropriate to Industry 4.0. This weakness is not only seen in the generation of skills among future workers, but also in the adaptation of the skills of today's workers. Third, the work has highlighted the importance of institutional structures that promote interconnection between regional innovation system actors in the early stages of development of an Industry 4.0-based business ecosystem. In this area, our research shows how in the Basque Country although an adequate level of public-private cooperation has been achieved, there is some weakness of the institutional structures that favour the exchange of knowledge of Industry 4.0 between companies. On the other hand, in Catalonia the institutional structures are weak both in the field of public-private cooperation and in the field of private cooperation. We also shown, that the specific public policy favouring the adoption of Industrial 4.0 strongly depends on the legitimacy and power games between the public and private agents of the innovation system. Finally, the work shows how the development of Industry 4.0 needs mechanisms that allow the development of a robust analytical knowledge base (science-driven knowledge) in these new technologies, but that needs to be connected to the synthetic knowledge base (engineering-driven knowledge). In both regions, this connection between the two knowledge bases is not observed, limiting the ability of the regional innovation system to adapt to Industry 4.0.

Notes

1. Spain is the second largest car manufacturer in Europe and the eighth largest in the world. In Spain there are 17 automotive manufacturer factories with more than 1,000 Tier 1, 2 and 3 suppliers. This strength implies that the automotive sector makes up 10% of Spanish GDP and 19% of exports.
2. The presence in Spain of 8 Airbus plants (3 in Andalusia and 5 in the area near Madrid) involves a network of 2,000 local suppliers.
3. Basque Industry 4.0 is a programme designed to promote Industry 4.0. technology transfer from R&D public centres to Basque firms in Industry 4.0. The budget of the programme in 2019 was € 2.55 million, with an average subsidy of € 80,000 for typically small R&D projects with no more than € 200,000 budget. Joint proposals of two or more firms are very rare in this programme.
4. After realizing this problem, the Catalan government started to merge slowly small technology centres into larger centres, such as Eurecat.

Disclosure statement

No potential conflict of interest was reported by the author(s).

ORCID

Francesco D. Sandulli ⓘ http://orcid.org/0000-0003-0831-9959

References

Ahedo, M. 2004. "Cluster Policy in the Basque Country (1991–2002): Constructing 'Industry–Government' Collaboration Through Cluster-Associations." *European Planning Studies* 12 (8): 1097–1113. doi:10.1080/0965431042000289232

Ahedo, M. 2018. "The Construction of Unbalanced Innovation Policies in Catalonia (Spain)." *International Journal of Innovation and Regional Development* 8 (2): 179–195. doi:10.1504/IJIRD.2018.092098

Albizu, E., M. Olazaran, C. Lavía, and B. Otero. 2011. "Relationships Between Vocational Training Centres and Industrial SMEs in the Basque Country: A Regional Innovation System Approach." *Intangible Capital* 7 (2): 329–355. doi:10.3926/ic.2011.v7n2.p329-355

Albors-Garrigós, J., C. Rincon-Diaz, and J. Igartua-Lopez. 2014. "Research Technology Organisations as Leaders of R&D Collaboration with SMEs: Role, Barriers and Facilitators." *Technology Analysis & Strategic Management* 26 (1): 37–53. doi:10.1080/09537325.2013.850159

Aragón, C., M. Aranguren, C. Iturrioz, and J. Wilson. 2014. "A Social Capital Approach for Network Policy Learning: the Case of an Established Cluster Initiative." *European Urban and Regional Studies* 21 (2): 128–145. doi:10.1177/0969776411434847

Aranguren, M., X. De La Maza, M. Parrilli, F. Vendrell-Herrero, and J. Wilson. 2014. "Nested Methodological Approaches for Cluster Policy Evaluation: An Application to the Basque Country." *Regional Studies* 48 (9): 1547–1562. doi:10.1080/00343404.2012.750423

Asheim, B., and L. Coenen. 2005. "Knowledge Bases and Regional Innovation Systems: Comparing Nordic Clusters." *Research Policy* 34 (8): 1173–1190. doi:10.1016/j.respol.2005.03.013

Autio, E., S. Nambisan, L. Thomas, and M. Wright. 2018. "Digital Affordances, Spatial Affordances, and the Genesis of Entrepreneurial Ecosystems." *Strategic Entrepreneurship Journal* 12 (1): 72–95. doi:10.1002/sej.1266

Bacaria, J., S. Borrás, and A. Fernández-Ribas. 2004. "The Changing Institutional Structure and Performance of the Catalan Innovation System." In *Regional Innovation Systems: The Role of Governance in a Globalized World*, edited by P. Cooke, M. Heidenreich, and H. J. Braczyk, 63–90. London: Routledge.

Borrás, S., and J. Jordana. 2016. "When Regional Innovation Policies Meet Policy Rationales and Evidence: a Plea for Policy Analysis." *European Planning Studies* 24 (12): 2133–2153. doi:10.1080/09654313.2016.1236074

Bostrom, N. 2007. "Technological Revolutions: Ethics and Policy in the Dark." In *Nanoscale: Issues and Perspectives for the Nano Century*, edited by N. Cameron, M. E. Mitchell, 129–152. Hoboken (New Jersey): John Wiley & Sons, Inc.

Castelo-Branco, I., F. Cruz-Jesus, and T. Oliveira. 2019. "Assessing Industry 4.0 Readiness in Manufacturing: Evidence for the European Union." *Computers in Industry* 107: 22–32. doi:10.1016/j.compind.2019.01.007

Coenen, L., J. Moodysson, and H. Martin. 2015. "Path Renewal in old Industrial Regions: Possibilities and Limitations for Regional Innovation Policy." *Regional Studies* 49 (5): 850–865. doi:10.1080/00343404.2014.979321

Feldman, M., J. Francis, and J. Bercovitz. 2005. "Creating a Cluster While Building a Firm: Entrepreneurs and the Formation of Industrial Clusters." *Regional Studies* 39 (1): 129–141. doi:10.1080/0034340052000320888

Flink, T., and D. Kaldewey. 2018. "The new Production of Legitimacy: STI Policy Discourses Beyond the Contract Metaphor." *Research Policy* 47 (1): 14–22. doi:10.1016/j.respol.2017.09.008

Fromhold-Eisebith, M., and G. Eisebith. 2005. "How to Institutionalize Innovative Clusters? Comparing Explicit top-Down and Implicit Bottom-up Approaches." *Research Policy* 34 (8): 1250–1268. doi:10.1016/j.respol.2005.02.008

Götz, M., and B. Jankowska. 2017. "Clusters and Industry 4.0–do They fit Together?" *European Planning Studies* 25 (9): 1633–1653. doi:10.1080/09654313.2017.1327037

Hernández-Lara, A., J. Moral-Martín, and I. Brunet-Icart. 2019. "Can Apprenticeships Contribute to Innovation in SMEs?" *The Case of Catalonia. International Journal of Training and Development* 23 (1): 7–26. doi:10.1111/ijtd.12144

Hickie, D. 2020. "Diversities of Innovation in Advanced Manufacturing – the Aerospace and Automotive Industries." *UCJC Business and Society Review (Formerly Known as Universia Business Review* 17 (1): 50–65.

Hilpert, U. 2020. "Divergent Opportunities of Innovation: When Time and Situation Matte." *UCJC Business & Society Review* 65: 18–35.

Ketels, C. 2004. "European Clusters." *Structural Change in Europe* 3: 1–5.

Konstantynova, A. 2017. "Basque Country Cluster Policy: the Road of 25 Years." *Regional Studies, Regional Science* 4 (1): 109–116. doi:10.1080/21681376.2017.1322528

Meniere, Y. Rudyk, J. Valdés. J. 2017. *Patents and the Fourth Industrial Revolution: The Inventions behind Digital Transformation.* Munich, Germany: European Patent Office.

Morgan, K. 2016. "Collective Entrepreneurship: The Basque Model of Innovation." *European Planning Studies* 24 (8): 1544–1560. doi:10.1080/09654313.2016.1151483

Morisson, A., and M. Doussineau. 2019. "Regional Innovation Governance and Place-Based Policies: Design, Implementation and Implications." *Regional Studies, Regional Science* 6 (1): 101–116. doi:10.1080/21681376.2019.1578257

Moso, M., and M. Olazaran. 2002. "Regional Technology Policy and the Emergence of an R&D System in the Basque Country." *The Journal of Technology Transfer* 27 (1): 61–75. doi:10.1023/A:1013148620724

Navarro, M., J. Martíns, S. Rodríguez, and A. Alonso. 2012. "Territorial Benchmarking Methodology: The Need to Identify Reference Regions." In *Innovation, Global Change and Territorial Resilience*, edited by P. Cooke, D. Parrilli, and J. L. Curbelo, 99–133. Cheltenham: Edward Elgar.

Nightingale, P. 2004. "Technological Capabilities, Invisible Infrastructure and the un-Social Construction of Predictability: the Overlooked Fixed Costs of Useful Research." *Research Policy* 33 (9): 1259–1284. doi:10.1016/j.respol.2004.08.008

North, K., N. Aramburu, O. Lorenzo, and A. Zubillaga. 2019. Digital Maturity and Growth of SMEs: A Survey of Firms in the Basque country (Spain). In *Proceedings, IFKAD Conference, Matera, 5-7 June.*

Orkestra. 2018. Competitiveness Report of the Basque Country 2017.

Orkestra. 2019. Competitiveness Report of the Basque Country 2018.

Reischauer, G. 2018. "Industry 4.0 as policy-driven discourse to institutionalize innovation systems in manufacturing." *Technological Forecasting and Social Change* 132: 26–33. http://doi.org/10.1016/j.techfore.2018.02.012

Riba Vilanova, M., and L. Leydesdorff. 2001. "Why Catalonia Cannot be Considered as a Regional Innovation System." *Scientometrics* 50 (2): 215–240. doi:10.1023/A:1010517505793

Rissola, G., and J. Sörvik. 2018. Digital Innovation Hubs in Smart Specialisation Strategies, EUR 29374 EN, Publications Office of the European Union, Luxembourg, 2018, ISBN 978-92-79-94828-2, doi:10.2760/475335, JRC113111.

Rodríguez Ferradas, M., J. Alfaro Tanco, and F. Sandulli. 2017. "Relevant Factors of Innovation Contests for SMEs." *Business Process Management Journal* 23 (6): 1196–1215. doi:10.1108/BPMJ-10-2016-0201

Rodríguez-Pose, A. 1999. "Innovation Prone and Innovation Averse Societies: Economic Performance in Europe." *Growth and Change* 30 (1): 75–105. doi:10.1111/0017-4815.00105

Scherrer, W. 2020. "How "General Purpose Technologies" Trigger Long Waves of Economic Development and Thereby Generate Diversities of Innovation." *UCJC Business and Society Review* 65: 36–49.

The growing inequalities in Italy – North/South – and the increasing dependency of the successful North upon German and French industries

Matteo Gaddi, Nadia Garbellini and Francesco Garibaldo

ABSTRACT
This paper analyses the impact of the adoption of 4.0 technologies and of the Italian Government Plan on the Italian industrial structure and on work organisation and workers' conditions. The Italian industrial structure is strongly unbalanced at a territorial level, because it is concentrated in the Northern country, while the South of the country is at a great disadvantage: the industrial and employment divide is therefore very evident. The adoption of 4.0 technologies and the Italian government's plan risk further aggravating this imbalance. Northern Italian industry, however, is also a cause for concern, as it is increasingly dependent on the supply chains of German, and to some extent French, industry. The consequences of the implementation of Industry 4.0 at the factory level on working conditions are negative for workers as it leads to an increase in work rhythms, an increase in workloads, a greater control over work performance and less autonomy for workers. In general, in fact, our research has shown a strong intertwining between 4.0 technologies and the organisational model of Lean Production, an intertwining aimed at increasing the exploitation of workers.

The Italian industrial structure

The Italian industrial structure is concentrated in the North with more than 50% of the manufacturing companies, almost 70% of the employees and about half of the inhabitants. We can consider the North of Italy made of two macro-regions:

North-West (Liguria, Lombardia, Piemonte, Valle d'Aosta) and North-East (Emilia-Romagna, Friuli-Venezia Giulia, Trentino-Alto Adige, Veneto); these two macro-regions have different kinds of specialization and different international links. The Centre of Italy (Toscana, Umbria and Marche) represent around 20% of companies, employees and inhabitants. The South (Abruzzo, Basilicata, Calabria, Campania, Molise, Puglia) and the Islands, considered together as the macro-region, named 'Il Mezzogiorno', represent a quarter of the companies, but only 14% of the employees and roughly 34 of the inhabitants. It is evident that in the South and Sicily and

Sardinia, the prevailing dimension of companies is small and very small. It is also apparent that the difference is not proportional to the inhabitants, and the main imbalance is related to the 'Mezzogiorno'.

Italy is divided along two axes. North and South, East and West. The nature of the divide is different. North and South are separated by different level of industrial development and by the availability and the level of public services and the territorial infrastructure. For instance, in the period 2015–2018, the economic growth, in Il Mezzogiorno, recovered not much more than 20% of the losses during the 2008–2014 crisis compared with the 85% of the Centre-North. In the tertiary sector, the South also recovered and surpassed pre-crisis levels. In terms of employment levels, the difference is striking; according to SVIMEZ (Svimez, 2015), it should be necessary to create three million jobs in the South to reach the rate of employment of the Centre-North, and avoiding that the existing unevenness will become deeper. In Italy, there is also a trend of differentiation between urban areas and the rest of the country. In terms of added value, the Centre-North represents 77.1% of Italy as a whole (2015). East and west are divided by the nature of the integration with the European industrial structure. Generally speaking, the East is more integrated with Germany and the Centre-east of Europe and the West more with France and the West part of Europe.[1]

We consider the engineering industry sectors as in Table 1 and their geographic distribution in Table 2.

It means that the importance of the main manufacturing sectors is different in the different areas.[2] Concerning Germany the main sectors in the engineering industry, for import-export trade, sectors 28, 25, 29 are followed by 23, 27, 30, 26 and 32 other manufactures. As from Table 2, sector 28 is the first sector in Lombardia, followed by Emilia-Romagna and Veneto; sector 25 Lombardia, followed by Veneto and Emilia-Romagna, sector 29 Piemonte, followed by Emilia-Romagna and Lombardia.[3] Considering industry as a whole, chemical products are very relevant with France the main sectors in the engineering industry, for import-export trade, are sectors 28, 29, 24. As from Appendix, sector 28 is the first sector in Lombardia, followed by Emilia-Romagna and Veneto; sector 29 Piemonte, followed by Emilia-Romagna and Lombardia, sector 24 Lombardia, followed by Veneto and Friuli-Venezia Giulia. Considering industry as a whole, Food and Clothing are very relevant.

We can divide the investigated companies into at least six macro-areas: automotive and motor vehicles, household appliances, general mechanics (motors, pumps, etc.), industrial

Table 1. Industrial sectors as to NACE Rev. 2 codes.

Sector	Proportion to total VA 2018, manufacturing, in descending order
25 – Fabricated metal products, except machinery and equipment	12.80
28 – Machinery and equipment n.e.c.	11.08
29 – Motor vehicles, trailers and semi-trailers	8.80
30 – Other Transport equipment	5.56
27 – Electrical equipment	4.17
24 – Basic metals	3.98
26 – Computer, electronic and optical products	2.78
33 – Repair and installation of machinery and equipment	2.61

Source: Authors' elaboration based on AIDA – Computerized Analysis of Italian Companies – database.

Table 2. Ranking orders of the regions as to their contribution to VA per sectors.

Ranking order Sectors	I	II	III	IV	Regions V	VI	VII	VIII
25	Lombardia	Veneto	Emilia-Romagna	Piemonte	Toscana	Marche	Friuli-Venezia Giulia	Campania
28	Lombardia	Emilia-Romagna	Veneto	Piemonte	Toscana	Friuli-Venezia Giulia	Trentino-Alto Adige	Marche
29	Piemonte	Emilia-Romagna	Lombardia	Veneto	Toscana	Abruzzo	Basilicata	Puglia
30	Lazio	Piemonte	Friuli-Venezia Giulia	Emilia-Romagna	Veneto	Toscana	Liguria	Campania
27	Lombardia	Veneto	Emilia-Romagna	Piemonte	Marche	Friuli-Venezia Giulia	Toscana	Lazio
24	Lombardia	Veneto	Friuli-Venezia Giulia	Emilia-Romagna	Piemonte	Trentino-Alto Adige	Umbria	Toscana
33	Lombardia	Emilia-Romagna	Veneto	Lazio	Piemonte	Campania	Toscana	Liguria

machinery and plants, steelmaking, installation and maintenance of plants (telecommunications, energy, etc.). In the automotive, household appliances and general mechanics sectors, mass production is organized according to the principles of Lean Production. In these sectors, production lends itself to being organized by implementing as far as possible the objectives of reducing the time assigned to workers, eliminating waste and achieving the tense flow. The mechanical engineering and plant construction machines and equipment is customized according to requirements what makes these frequently to be unique pieces (one piece). In this case, companies try to standardize the production of parts and components (generally mechanical departments), while they have more difficulty in standardizing assembly phases. In steelmaking the duration of chemical–physical processes cannot be compressed; also, in this case, the companies' attempt is to recover the reduction of time in the activities downstream of the melting and casting processes, i.e. in the rolling and metalworking phases. In-plant installation and maintenance, despite the high variability of the interventions to be carried out, the companies attempt to standardize every kind of intervention; to assign a realization time and to control it through remote control instruments; including geolocation instruments for workers.

During recent decades there was a transformation of the industrial structure of the North of Italy. Until the 1980s and 1990s, the backbone of the North was the so-called industrial triangle, whose vertexes were Torino, Milano, Genova. Now the more realistic description, designed by the North-East Foundation,[4] is a pentagon containing the North-East territories, namely five regions, from west to East: Lombardia, Emilia-Romagna, Trentino-Alto Adige, Veneto, Friuli-Venezia Giulia. Then, the North-East Foundation selected some European Regions as a benchmark to rate the Italian Regions. The Benchmark regions are Baden-Württemberg, Bayern, Auvergne-Rhône-Alpes, Alsace-Champagne-Ardenne-Lorraine, Este (ES), Western Austria e Noreste (ES). The Pentagon as some indicators – such as GDP/person, unemployment rate, NEETs (Not in Education, Employment or Training) rate, and market openness – is close to the benchmark. Southern companies are at a disadvantage compared with the ones in East of Europe. This disadvantage is due to the level of wages and taxation. Besides, in general terms in Europe, the Eastern countries are converging, as to the incomes, and the Southern are diverging. These kinds of difference within Italy are, in some way a micro-representation of the European Union internal divides.

Besides, Italy is in a demographic decline which also differs clearly by regions. In the South, the decline is much worst. SVIMEZ forecasts, in 2065, a 15% decrease of the working-age population in the Centre-North (−3.9 million) and a 40% in the South (−5.2 million). It means that with current levels of employment, productivity and migration balance, Italy will lose almost one-quarter of GDP, but in the South, it is more than one-third.

The relevance of the societal structures

It is a key point in our hypothesis that Industry 4.0 development depends on a social basis and its disparity in the different parts of Italy; therefore, some social indicators are highly relevant: demography and education. Because of the emigrations, the Italian population will decline in the centre-north. The decline will be limited, thanks to the immigration from abroad and the South. The migration from the South mainly includes young people with a university degree, and most of them will never return. As to education according to the Excelsior report on the MEDIUM-TERM EMPLOYMENT AND PRO-FESSIONAL NEEDS FORECAST (2019–2023),[5] the employment needs of the companies in the 'mechatronics and robotics' sector may concern between 68,000 and 86,000 workers, also over the five years. The different figures depend on different hypothesis on the growth rate of the Italian economy

> Namely, mechatronics is the manufacturing sector most affected by new production methods. We mean, the ones synthetically summarised with the term 'Industry 4.0'. Besides, IT services and advanced services, of course, play a leading role. These sectors will require a significant number of professionals linked to the transformation of production.

> The evolution of the '4.0' productions – in particular in mechatronics-robotics, but also many other sectors – brings with it a growth of the professions which determines an ever-lesser ability of the current classifications to identify professional figures correctly. Just think of the growing difficulty of distinguishing between 'technicians' and 'workers' and between skilled and unskilled workers. The request for digital and other skills to all the employed, even the less qualified ones, entails problems of defining the professional competences that cannot be easily solved.

Focusing on the problem of orienting and programming the training of the new employed, what has just been detected opens up challenges utterly original for the training operators themselves. But the regional disparities are evident also in the education performances.

The distribution of educational qualifications for those aged 15 or over is as follows:

	Educational qualifications			
Age	Primary school or no qualifications	Secondary level	Vocational qualification 2–3 years	Upper secondary level 5–6 years
15 and more	8,822,000	16,800,000	2,835,000	15,926,000
%	16.96	32.29	5.45	30.61
	Of these 40% South and Islands			Of these 67.29% in Centre-North

The last class of upper secondary education in Italy, in 2019, has the following distribution of mathematics proficiency: 40% below standards, and 60% in standards or above, of which 10% at the highest level. As to regional differences, in Center-North, Umbria is in the national average, Lazio below, and all the others are above the norm. The North-West between 70% and 80% of values in the standard or higher, with between 25% and 35% at the maximum level. The North-East, between 65% and 80% of the standard or higher, with between 25% and 35% at the maximum level. The South + Islands range from 45% to 65% below the norm and vice versa, with 10–15% at the maximum level. The worst case is that of Calabria with 65% and 35% and 7% at the maximum level. In English the national average is 50–50 for reading and 65% below and 35% in standard or higher for listening. The regions above the national average, for reading, are in Center-North except for Lazio and Tuscany. The peaks over 60% are represented by the North, with excellence, over 70% of Trentino-Alto Adige. The regions above the national average, as to listening, are in Center-North with the exceptions of Lazio and Umbria. The peaks over 60% are represented by only Trentino-Alto Adige.

According to Eurostat: In 2018, more than two-fifths (40.7%) of the EU-28 population aged 30–34 years possessed a tertiary level of education; as such, the ET 2020 benchmark was attained with two years to spare. The EU-28's tertiary educational attainment among people aged 30–34 years rose by 9.6 pp between 2008 and 2018, and by 0.8 pp between 2017 and 2018. Across the EU Member States, attainment levels in 2018 ranged from a low of 24.6% in Romania and 27.8% in Italy to cover more than half of this subpopulation in Sweden (52.0%), Luxembourg (56.2%), Ireland (56.3%), Cyprus (57.1%) and Lithuania (57.6%). But if we consider the regional differences in Italy, the North and Lazio are in between 30% and 40%, the rest of Italy in between 20% and 30%. The share of individuals aged 18–24 years who have at most a lower secondary level of educational attainment (ISCED levels 0–2) and who were not engaged in any further education and training (during the four weeks preceding the labour force survey (LFS)), in 2018 stood at 10.6%. Across the EU Member States, the proportion of early leavers from education and training ranged from 3.3% in Croatia up to 17.9% in Spain: this distribution was skewed insofar as just nine Member States recorded shares above the EU-28 average.

In comparison, 19 Member States had lower shares – 17 of which recorded percentages of early leavers from education and training that were below the 10% policy target. Most of the Italian Regions are in between 10% and 15%, but the South I in the upper range starting from 15% and reaching more than 20%. These dire performances in some Italian Regions emphasize the contrast with what companies consider a must in the new industrial revolution. Consequently, the Northern regions provide much better educated and skilled labour. Thus, there is a current and constantly increasing high level of competences of the regional societies. This allows widely for the application of opportunities associated with Industry 4.0 in the Northern regions and indicates a growing inequality when compared with the Centre or the South.

The situation of Italian manufacturing industry

The Italian Industry 4.0 plan was born following the German one. It depends on the deep integration of a large part of the industrial system of northern Italy with the German one. SMEs are an important element of national and regional strategies. Still, in the case of Italy, for the main industrial sectors, there is the risk of companies becoming increasingly dependent on OEMs of different value chains. The OEMs are often located in other countries. This is particularly true for companies located in the North-West and the North-East of Italy.

There are some distinctive aspects of the Italian industrial structure, also stressed by the government, which make it mainly to be a subject that is affected by Industry 4.0. Particularly in mechanical engineering which is oriented in industrial automation and components (mechanics and mechatronics), SMEs are clustered in industrial districts (spread throughout Central and Northern Italy) which would become more and more integrated with a reduction of the distance, within the value chains, between suppliers and subcontractors. Besides, depending on the sector, SMEs have different challenges to cope with. There is a widespread idea in the business community about a retrofitting strategy that should be appropriate for many SMEs. But this is not suitable for SMEs which depend on global or EU value chain with OEMs still on the edge of the process. As a matter of fact, according to our research, value chains is very different when compared with the previous decade. Today, the first level suppliers are strongly regulated, by OEMs, as too many dimensions and namely concerning the nature and the pace of their investments to be fit with the new standards.

This inhibits some severe problems to be dealt with on the firm level. Finally, what is worrying for the labour organizations, the Minister endorsed the need for making industrial relations more flexible by decentralizing bargaining activities to the level of the single firms. Besides, it stressed the necessity of closely linking salary adjustments and corporate productivity following a model which has been strongly supported by the latest governments and CONFINDUSTRIA (the General Confederation of Italian Industry) in recent years. This relationship between wages in regions and firms which participate in innovation based on Industry 4.0 development and productivity has proved to be very detrimental to workers: first of all, because the term 'productivity' has only meant higher business profitability. Besides, 'productivity' bargaining was based on indicators over which workers and trade unions could not exercise any control. From this point of view, the regional differences concern only the spread of second-tier bargaining, which is concentrated for about 80% of the contracts stipulated in Northern Italy (about 39% for both the North-West and the North-East).

The 2016 CNEL-ISTAT Report on 'Productivity, Structure and Performance of Exporting Enterprises, Labour Market and Supplementary Bargaining' (CNEL-ISTAT, 2016), provides specific data on the disparities between the North and the South. In the manufacturing sector, 36.8% of enterprises have some form of second-level bargaining. Still, collective bargaining exists only in the 25.1% of enterprises, while in the remaining part (11%) exists only individual negotiation. Therefore, most companies do not have second-level bargaining. The highest concentration of decentralized contracts is located in the North-East (38.9% of the total) and the North-West (35%): in these two areas. Therefore, more than 70% of decentralized bargaining is concentrated.

The government plans: the design of the government program in the light of existing uneven regional situations

The Italian government in 2015 gave its opinion, in the person of the Minister of Economic Development, Carlo Calenda, during a parliamentary hearing at the Chamber of Deputies. According to the Minister, digitalization improves the competitiveness of the Italian manufacturing sector, starting from those production chains mainly based on SMEs. The government's plan opened a phase of the progressive introduction of manufacturing digitization processes and a growing interest and discussion on the consequences of this choice. The debate involved both the research world and the business organizations and the trade unions. The Minister ruled out the possibility that Italian industrial policies can foster the development of vertical chains, giving priority instead to a horizontal approach based on innovation, internationalization and recourse to the capital market.

The concept is supporting the industrial structure in its existing configuration, without any explicit objective of reducing the territorial and the dimensional imbalances of this structure.[6] The government's plan (Governo Italiano, 2016) in September 2016 (National Plan Industry 4.0, later renamed Impresa [firm] 4.0) designed such relevant incentives to stimulate the spread of Industry 4.0 that the business world said it would slow down the planned investments.[7] This plan includes several areas of intervention. The first one is intended to provide for investment in innovation and legal incentives (laws for machinery modernization, patent box, tax credits on R&D, etc.). The second area concerns investments in technologies (connectivity infrastructure, reduction of SMEs' digital divide, improvement of STEM skills). The third area concerns interoperability and communication standards to foster production processes and business models based on IoT. The fourth aims at developing corporate finance to support companies' investments for Industry 4.0.

This plan has many critical elements. First of all, the entirety of the plan's objectives concerns only the firm-level. The main goals are: (1)Greater flexibility, which would allow taking advantage of economies of scale even when producing small batches; (2) shortening the time necessary for prototypes to be switched-over to series production; (3)decreasing set-up time, mistakes and machines stop, which would enhance productivity; (4) the introduction of sensors monitoring production in real-time, which would increase quality. These objectives should be consistent with the imperative of competitiveness. Competitiveness should be achieved thanks to the advantages offered by the Internet. Social and labour issues are not considered, but as to the functional ones as skills.

The attempt to provide the infrastructure necessary already shows the plan's first criticality because the realization of ultra-broadband is hugely behind schedule. Besides, the basic idea is that public commitment may stimulate private investments in goods and technologies which are connected to Industry 4.0. However, it is not clear whether such investments would generate production and employment in Italy or abroad. In other words, are sensors, devices, robots, hardware, software, and so ongoing to be produced in Italy or to be imported? If the latter is the case, besides a worsening of the Italian trade balance, the Government's Plan is going to generate employment abroad rather than in Italy and to support regional development in Italy.

The list of investment goods that can benefit from tax advantages is extensive. Companies could benefit already when buying capital goods connected with some form of ICT technology. But there are at least two constraints: one systemic, the uneven geographic distribution of companies and the latter also at the company level the education features. In the absence of any prospect of intervention in the Italian industrial structure, the advantages benefited the most structured companies, with the risk that this could increase social and regional inequalities. Consequently, when it comes to necessary investments, it turns out clearly that there is a sharp divide between the North and the other Italian regions.

Moreover, the plan hardly mentions the absolute predominance of small enterprises. This represents a chronic weakness of the Italian industrial system. They are usually part of more complex production chains, whose head is located abroad, often in Germany. The government's plan inhibits a risk of digitalizing a set of production chains led by German companies, whose leadership on the whole supply chain would be even reinforced. As a consequence, when applying opportunities for Industry 4.0, also the most industrialized and modernized regions risk further deepening their dependence on the international industry. In this way, Northern companies supplying components to the German industry would be increasingly integrated into the production structure of Germany, but in an increasingly subordinate position. The economic success is related to integration into an innovative system which allows little control over the own position. Companies from the South, on the other hand, risk being further marginalized. In both cases, therefore, a general weakening of the Italian industry as a whole would result.

As a consequence, such innovation again makes regional development more uneven. The plan, by its admission, adopts a horizontal approach to industrial policies, with the explicitly stated aim of avoiding the vertical approach. For instance, the key topic of the future of the automobile sector and its industrial ecosystem is not considered, but this is a critical issue. In fact,

> the direct and indirect automotive industrial sector has 5,700 companies, 258,700 employees, which generate a turnover of over 100 billion euros equal to 5.9% of GDP and gross fixed investments of 3.9 billion euros, 13% of the investments of the Italian manufacture (…) The direct employees of the industrial sector … are 162,000, of which 67,000 employed in the production of motor vehicles and their engines.

Of no less importance is the tertiary sector, which 'has 225 thousand companies, 942 thousand employees generating a turnover of 230 billion euros'. Besides, 'the capacity of the sector for activating employment has a multiplier equal to 3, that is, ten employed in automotive companies in the industrial phase, they support 20 additional employees in the economy'. As regards the components, there are 2207 companies with 158,700 employees and a 2018 turnover of 49.345 billion euros. 37% of this turnover depends on FCA, but the others depend on the European value chains.

The plan lacks any reference to labour, but in terms of training and skills development. This is particularly important for the situation of regions in Northern Italy. One example suffices: Industry 4.0 generates new jobs (new sectors, products, services) and, at the same time, destroy jobs (due to automation, robots, etc.). Estimating job balance at the single firm or sector level appears to be feasible. Moreover, it would be worth trying to understand how Industry 4.0 changes workers' status (new and more flexible forms of employment; the dichotomy between employment and self-employment, etc.) and working conditions (working time, safety at work, etc.). How will companies and territories comply with the new paradigm? Which new skills become necessary? How will working tasks and processes change? How will working performance take place, and be monitored by each firm? All these are questions which the government's plan does not even mention. This list of questions should be made explicit as there has been no prior assessment of how the work would have changed, especially for work organization and production. As usual, the only point of view adopted is the enterprises' one.

As a general assessment, also the government programme did not consider the regional disparities so that the digitalization process run the risk of deepening these disparities. For the geographical distribution North-West, North-East and Centre have made greater use of the incentives of super-depreciation and hyper-depreciation with the South left behind. The gap between the South and the rest of Italy, in the utilization of the incentives, is less relevant in the case of the use of tax credit on R&D and the Nuova Sabatini's incentives.

In 2017 (Ministero dello Sviluppo Economico, 2018) about 80% of large companies used the incentives of the government plan, while only 20% of the small companies did so. Although the different use of incentives was reduced during the following year (2018), still, small companies continue to be very different from large ones. Most of the incentives were used to fund the purchase of technologies such as software and the IoT (i.e. technologies directly linked to connected capital goods.) A significant spread of these technologies has been found in engineering industries (automotive, white domestic appliances, steel and iron production, industrial machinery, transport equipment, etc.), chemical, telecommunications and logistics companies. Whereas, in these industries, companies of 250 employees or more stand for only 0.3% of the total number of enterprises. Since these larger firms are concentrated in the Centre-North of Italy, when realizing these plans and programmes, there is a risk of intensifying the current uneven development.

On the effectiveness of the plan: the increasing regional inequality

A qualitative survey conducted by ISTAT in 2017 (Istat, 2018) on a representative sample of manufacturing companies provided some indications on the perception by the business world on the effectiveness of 'Impresa 4.0'. For 62.1% of manufacturing enterprises, super-depreciation played a 'very' or 'fairly' significant role in the decision to invest. As to the managers' appreciation, 53.0% of medium-sized enterprises, 57.6% of large enterprises, and 34.2% of enterprises, with less than 50 employees, considered hyper-depreciation a relevant measure for their investments' decisions. The appreciation values ranged from 57.3% of small to 66.9% of medium-sized enterprises. The tax credit for R&D expenditure was considered favourable by more than 40% of manufacturing companies, a percentage that is close to 50% in larger companies. The financial benefits provided by the 'Nuova Sabatini', an instrument introduced to encourage investment in the capital goods of smaller companies, were considered necessary by 35.2% of small and 28.9% of medium-sized companies.

From a sectoral point of view, super-depreciation was considered positive by at least half of the companies in all sectors (except for clothing and other transport equipment). Hyper-depreciation, instead, was relatively more important for companies operating in the electrical equipment (58.9%), rubber and plastic (57.7%), metallurgy (55.8%), electronics and machinery (53.6% in both cases) sectors. The R&D tax credit was considered useful mainly by companies in the automotive (69.8%) and other transport equipment (60.0%) sectors. Concerning investment plans for 2018, almost 46% of enterprises state that it expects to invest in software, nearly a third (31.9%) in technologies machine-to-machine or Internet of things. Only 27% invest in connection with the Internet, high speed (cloud, mobile, big data, etc.) and computer security, to the extent that the directly proportional to the size of the business. The more innovation related to Industry 4.0 concentrates and suits the particular situations of Northern Italy, the more uneven regional development is intensified across the regions.

The new Conte government in the programmatic declarations again has put at the centre the project of an extensive digitalization of companies with particular attention to SMEs. The 2020 Budget Law (Law no. 160/2019) introduced a new tax credit for investments in new capital goods, replacing the extension of the super-amortization and hyper-amortization, which remain applicable to investments in capital goods made until 31 December 2019. The beneficiaries of the tax credit are companies that from 1 January 2020 carry out investments in new capital goods for production facilities located in the territory of the State. Companies resident in Italy, including permanent establishments of non-residents (i.e. foreign-owned companies), are eligible for the tax credit, regardless of its legal form, economic sector, or size. With the 2020 Budget Law the new tax credit framework for investments in research and development, ecological transition, technological innovation 4.0 and other innovative activities to support the competitiveness of companies has been introduced. It also extended the benefit of the tax credit for training 4.0 to 2020.

The patterns of change and the consequences for employees[8]

In addition to new technologies and new management models, there are new service contents, concerning data collection and processing. These new services come from the OEMs and their service firms. This high-level service again is realized in the North or abroad. According to the managers' interviews, there are two different paths of innovation based on Industry 4.0: the smart factory, the smart products. There are also companies affording both, but frequently even in these cases, a path is the prevailing one. The way toward producing smart products is based on achieving a new business model based on the integration of manufacturing and services in smart products. In some sectors, the smart component of the product is more relevant, regarding the value and competitive edge, than the manufactured part. Besides, smart products usually include a basket of services for the client – as, for instance, the possibility of predictive maintenance – allow the collection of data – an enormous amount of data – by customers and clients to feed AI-based innovation shortly.

When summarizing these main findings (Gaddi et al., 2017), there are clear implications for a smart factory. It requires a reorganization along the lines of the lean production, of the tense flow scheme and production flexibility. In conclusion, this eliminates all non-value-added activities and introduces feedback mechanisms to align production with demand. These objectives were first pursued mainly with organizational measures. Now Industry 4.0 technologies allow this to be accomplished primarily by technology but presuppose previous organizational changes. In all, there have been critical structural changes – new plant layouts – both in IT and in smart tools. The adoption of the new MES (Manufacturing Execution Systems) management systems, i.e. systems that unlike the old MRP (Material Requirement Planning), support in an integrated way the production both in the management of production operations and in those supporting production.

The companies best prepared to accept the technologies of Industry 4.0 seem to be those that had already implemented organizational innovations in the direction of lean production. Besides, companies owned by multinationals, mainly concentrated in Northern Italy, or with close ties to German industry (also in this case located in Northern Italy) have shown a higher propensity to Industry 4.0. There is, therefore, no standard formula for the application of Industry 4.0. The forms depend on the nature of the strategic problem of the product/market relationship that every company must solve. However, there are in all some common elements. Elimination of non-value-added activities more effectively and pervasively than before. Achieving this goal through technical devices changes the relationship between the hierarchy and the forms of the work command. Everything becomes less mediated by the social contact and more by the apparent objectivity of a relationship mediated by technology. Besides, the supply network is integrated directly – at least at the first level – in production management. There are currently two integration profiles: (i) An operational integration mainly through the adoption of the parent company's standards by the suppliers; and (ii) In addition to operational integration, real co-design relationships develop. There is a previously unthinkable control not only of performance as a result but also of its execution moment by moment and in real-time. This represents a shift – with the ideal model of the tense flow – towards a flexible and continuous alignment between market and production through the use of cyber-physical systems able in different measures to self-settle concerning market demand. The ideal model is batch production one.

The social side

The previous points involve a significant intensification of the work that must be evaluated case by case in its effects, also as to the psychological factors. Future employment consequences must also be assessed. In our research, although concerns have been expressed, there does not seem to be significant consequences at the moment. In regions where Industry 4.0 applies, there are quite a number of societal issues concerned. Participation with changes in progress requires a multi-level assessment. On the first level, the participation of workers must be evaluated and only in some cases, there are practices of involvement of individuals, but of a functional nature. Functional participation is achieved through both initial and ongoing training on ongoing changes. On the second level, that of the company and the provincial unions. In all the cases analysed, the union and the Trade unions' delegates (RSU) are at least informed. There are significant cases of proactive practices of the RSU. Even when they do not realize real framework agreements can intervene promptly to protect workers and workers involved. At the third level, there is the negotiated participation with the presence of real joint commissions. It happened only in a few cases.

In regions of the North, where industrial structures and skilled labour allow for participation in Industry 4.0 work–life may change significantly. The consequences on work, caused by Industry 4.0 transformation, apply both to manufacturers of final goods and manufacturers of parts and components intended for supply. Industry 4.0 enables a reduction of working times in a way adverse to workers. There is, indeed, an intensification of the pace of work and a decrease of time available for each operation. This modification was not due to the mere introduction of new technologies, but rather to the implementation of new business models, strictly determined by market conditions. Technologies supported these new business models, making a different organization of labour possible by reducing operation times. Order fulfilment times became stricter, strongly influencing working times and schedules. For this reason, a series of software tools acquired greater relevance. Such changes in work–life are related to societies and social life in Northern Italian regions.

The intensification of workloads was made possible by technologies that can track the start and end of every single operation. Data concerning operation times are recorded, collected and monitored thanks to computer-based systems. Moreover, Companies introduced devices for remote control of plants and equipment and, therefore, of the corresponding workers' performance. This control can be realized by matching barcodes associated with workers (ID badge), the machines they operate, the batch which is being produced, and the specific components under process. Machines also generate data about production volumes and downtimes (breakdowns, set up, controls, lack of materials, etc.).

In the vast majority of cases, Companies unilaterally define work schedules and working times. This is often the pre-condition to implement labour organization models such as Lean Production, Just in Time; Just in Sequence, WCM etc. These labour organization models – supported by Industry 4.0 technologies – enable: (i) compliance of supplies deliveries to planning defined by the company at the head of a production chain (OEMs); (ii) the synchronization of production stages (internal and external); and (iii) the management of the high degree of variability of workloads and product mix. Work orders have a barcode that embeds cycle times, often based on machine times. In this way, working times are presented by Companies as 'objective', and as such, not subject to bargaining. Finally, all data are immediately uploaded, collected and analysed by ERP/MES. Companies, via these control systems, can compare internal costs with costs charged by potential external suppliers. They create competition between internal and external workers and put much pressure on their employees.

While there is much reorganization and different skills required, these Northern Italian regions are also confronted with a contrary societal development. The utilization of more advanced tools and machines (connected devices, smart devices, etc.), therefore, does not imply higher-level skills for workers. On the contrary, tasks might even result simpler and poorer. Despite the rhetoric of Industry 4.0 about the upskilling of workers, we found that companies exclude workers more and more. This exclusion concerns all aspects related to production process information. The fact that workers are not aware of the working of the machine is a potential additional source of alienation: information, data, scripts by which the system works are entirely unknown. The cycle time of the operations performed by the machines is embedded in the software programs that govern their process. The worker's performance, in this way, is entirely dictated and constrained by the operation of the machine.

The constraint on the worker is further worsened by the fact that thanks to the networking of Industry 4.0, the machine operation programs (and therefore their cycle times) can be continuously re-determined based on data coming from the field (shop floor). The elaboration of software and scripts is of competence of planning and engineering departments. In some cases, this service is not even provided internally, but by the companies which supplied equipment, which is in charge of software updates, maintenance, etc. Programming requires computer science skills. It is also true that workers could well be involved in the discussion concerning software goals (machines' performance, modes of operation, times, etc.). Consequently, Industry 4.0 is more than technological innovation; it also changes the regional societies. In smart factories which are more frequently steered from outside management changes and there is also a tendency towards de-skilling of work. The rhythm of the new machines and the organization of manufacturing changes work life. The changing demands for skills and education attract employees from outside with new and modern skills, who often migrate their coming from the South. Finally, the regional societies in Northern regions become even more different from the South, along with a constant modernization of industry and economy.

Conclusions

Among Italian regions, processes associated with Industry 4.0 are indicating fundamental changes in regional industrial structures and social life. While the Northern regions can both benefit from these processes and enter an intensified dependency, the other regions and particularly the South will have little relation to these changes. This meets a situation, when Italy during the recent decade lost about a quarter of its industrial production, still has not fully recovered, and the GDP per capita is still reduced by 8%. In the Southern regions, it leads to a creeping process of deindustrialization and degradation of the productive matrix, associated with two digits losses GDP per capita. The integration of Italy into the EU industrial structure is very different for the northern regions and the rest of Italy. The Italian northern industrial structure is fully integrated into the leading European value chains. The drive to Industry 4.0, without specific policies to overcome these disequilibria, will confirm this divide. This is the more critical issue also in the political arena. Thus, when Northern regions will be integrated successfully into the European industrial structure by applying Industry 4.0 this will also introduce an increasing dependency from supply chains in France and Germany.

Although in Italy and for its regions SMEs play an outstanding role, but digitalizing Italian SMEs is now a key political issue. Similarly, upgrading of skills of employees and the continuous training of the new entrants has no priority. In light of this situation, trade unions and employer associations aim to balance this problem by providing training material and training packets for their associates and officials. The Association of the Engineering sector' employers, as an example, sponsored a book to explain what digitalizing manufacturing is about and to deliver pieces of advice for SMEs (e.g. by suggesting small firms to avoid a full implementation of all technologies at once). There are already some new good cases of the company's agreements to afford the chief complaints of the employees and to smooth the way to the digitalization. Industry 4.0 programme should be more aware of the problems associated with this innovation, also taking the North–South divide into consideration – in favour of the South. A successful introduction and application of Industry 4.0 technologies and equipment in the North will increase the uneven development in the North and South of the country as well as contributing to a continuation of young well-educated people to the North. Thus, innovation may contribute to continuing and intensifying of uneven regional development by industry and society – this it may have an impact on further deindustrialization of Southern regions.

Notes

1. Our research sample is made up of engineering industry companies (industrial sector NACE Rev. 2 codes 24, 25, 26, 27, 28, 29, 30, 33), and in these cases the distribution of the companies is more unbalanced because more than 65% are in the north of Italy with roughly 73% of the employees.
2. It is also evident the existence of a different specialization between the North-West and the North-East, as well as between North, Center and South (see Table 2). In the first three positions the only Region of the South is Lazio and none of the Centre. The best positioning of the Centre is in the fifth position in four sectors out of seven. Campania, besides, is in the sixth position for sector 33 and the eighth position for sectors 25 and 30.
3. World Input-Output Database.

4. The selection of these Regions is based on an indicator ISES – index of economic and social development – worked out from the North-East Foundation whose members are the employers' associations of Veneto, Friuli-Venezia Giulia and Trento. The index compares the Italian provinces with each other, on this ground it is possible to compare the different Regions. They perform much better than other Italian Regions and better than the North-West.

5. https://excelsior.unioncamere.net/index.php?option=com_content&view=article&id=364: previsioni-dei-fabbisogni-occupazionali-e-professionali-in-italia-a-medio-termine-2019-2023&catid=108&Itemid=1698.

6. Besides, the Minister didn't make available the analytical data on the utilisation of the funds, so we can say for sure that there was not a convergence process, but we cannot produce any analytical data set beyond the ones here listed.

7. This confirmed the concern of our research team about the fact that the only interest of Italian companies was to benefit from generous tax incentives and economic contributions.

8. We started a field research on the introduction of industry 4.0 in the metalworker's sector.
   Our sample, available in a report available in English, is made of 30 companies – 3 in Piemonte, 6 in Lombardia, 15 in Veneto, 6 in Emilia – Romagna – ranging from metal casting companies to an automaker, and a sample of different producers of machine tools. This is a work in progress, and we have more case studies in some reports, available only in Italian.
   These companies represent different sectors according to the NACE codes. As it is easy to check this sample is very coherent with the actual distribution of companies involved in the Industry 4.0 project funded by the Italian state. We carried out the research interviewing both the managers of the companies and a sample of workers and their delegates.

9. The result was obtained as the ratio between the sum of the value added produced by all the enterprises in each sector at regional (numerator) and national (denominator) level.

Disclosure statement

No potential conflict of interest was reported by the author(s).

References

CNEL-ISTAT. 2016. "Produttività, struttura e performance delle imprese esportatrici, mercato del lavoro e contrattazione integrativa", Rome.

Gaddi, M., N. Garbellini, and F. Garibaldo. 2017. "Industry 4.0. and Its Consequences for Work and Labour." http://www.fondazionesabattini.it/ricerche-1/ricerca-europea-industria-4-0.

Governo Italiano. 2016. "Piano Nazionale Industria 4.0. Investimenti, produttività, innnovazione." Rome.

ISTAT. 2018. "Rapporto sulla competitività dei settori produttivi, Edizione." Rome.

Ministero dello Sviluppo Economico. 2018. "La diffusione delle imprese 4.0 e le politiche: le evidenze." Rome.

Svimez, 2015. "Rapporto Svimez 2015 sull'economia del Mezzogiorno." Rome.

Appendix[9]

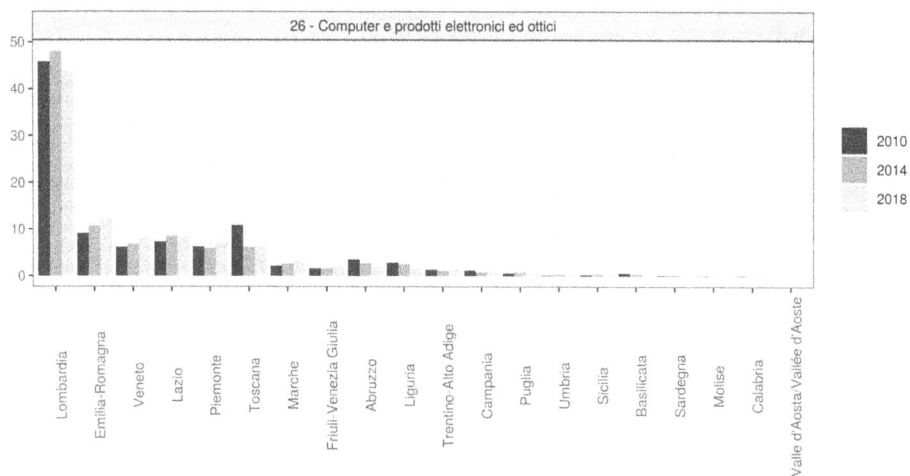

26 - Computer e prodotti elettronici ed ottici

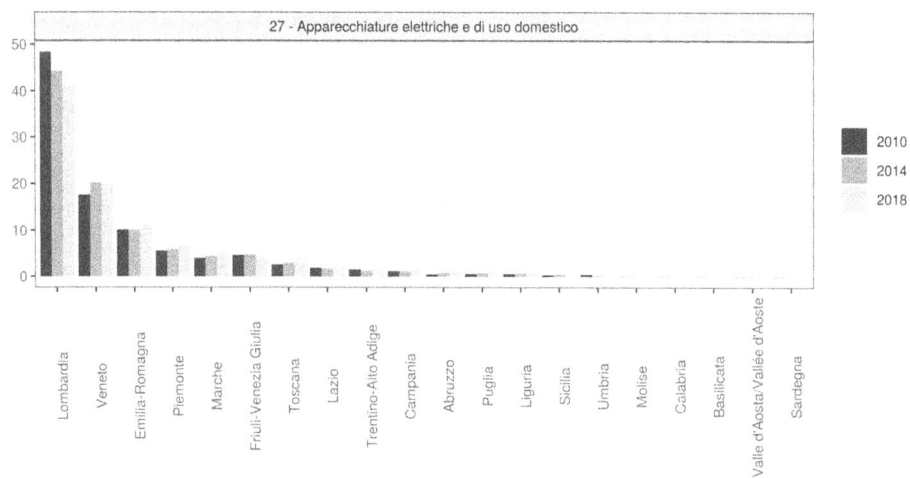

27 - Apparecchiature elettriche e di uso domestico

[9]The result was obtained as the ratio between the sum of the value added produced by all the enterprises in each sector at regional (numerator) and national (denominator) level.

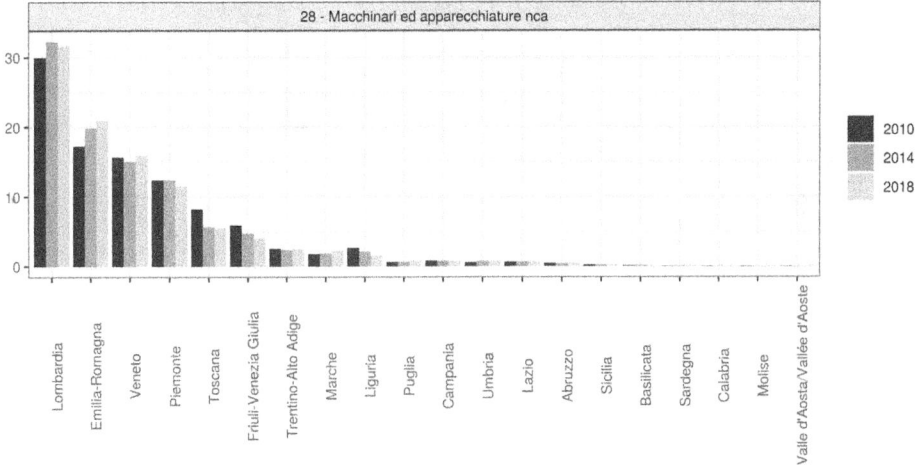

28 - Macchinari ed apparecchiature nca

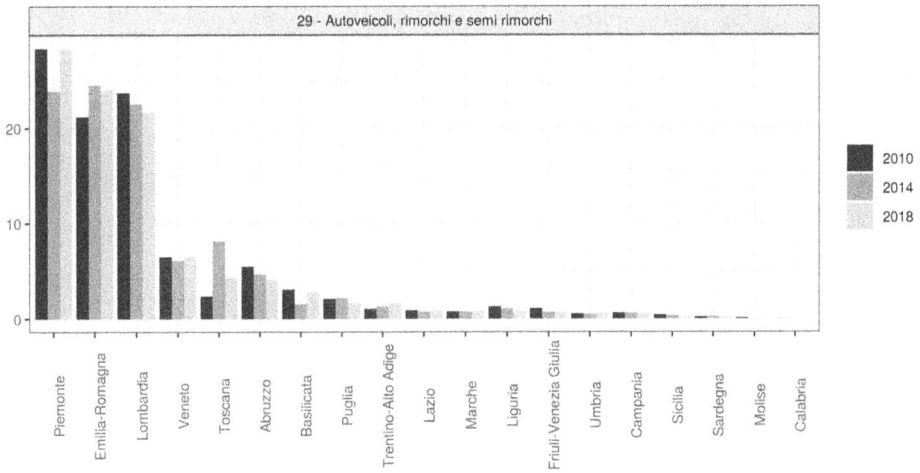

29 - Autoveicoli, rimorchi e semi rimorchi

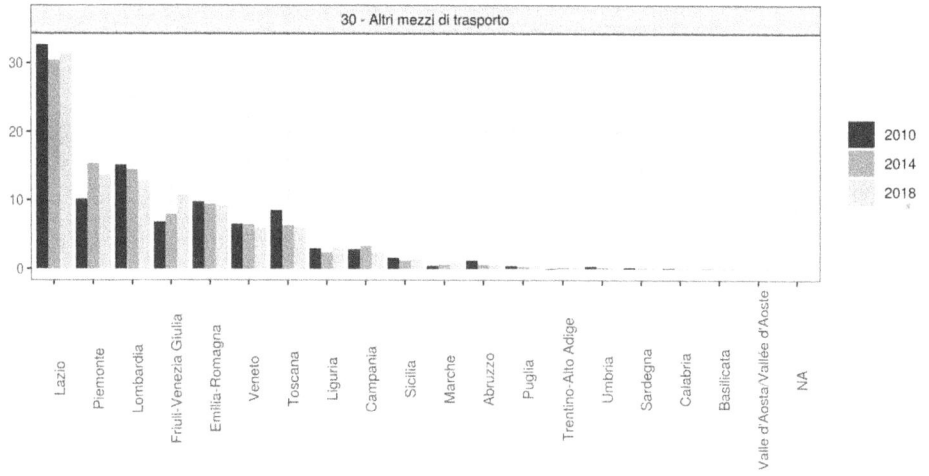

30 - Altri mezzi di trasporto

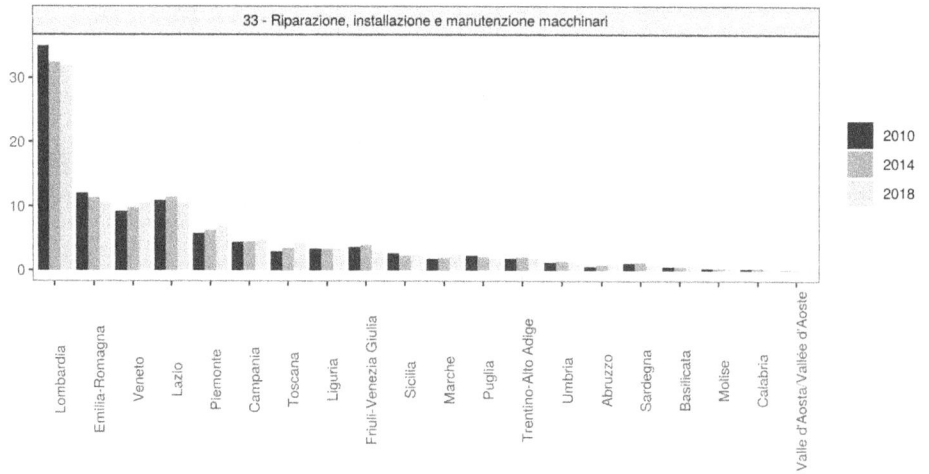

33 - Riparazione, installazione e manutenzione macchinari

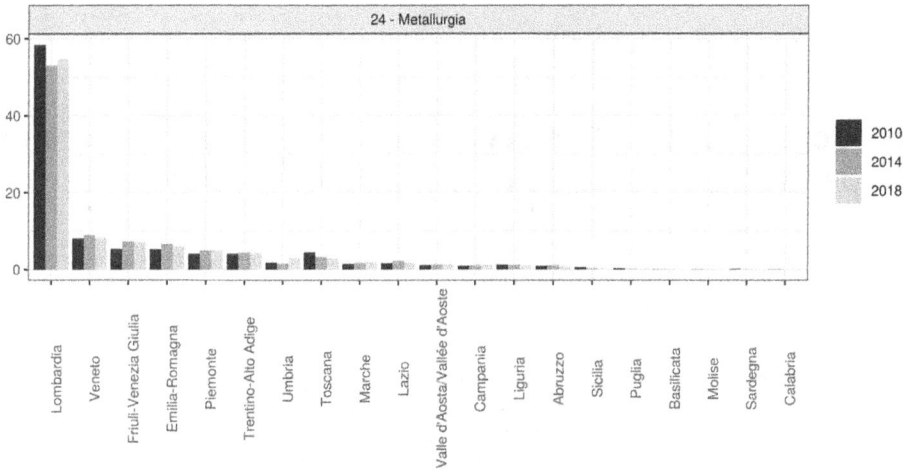
24 - Metallurgia

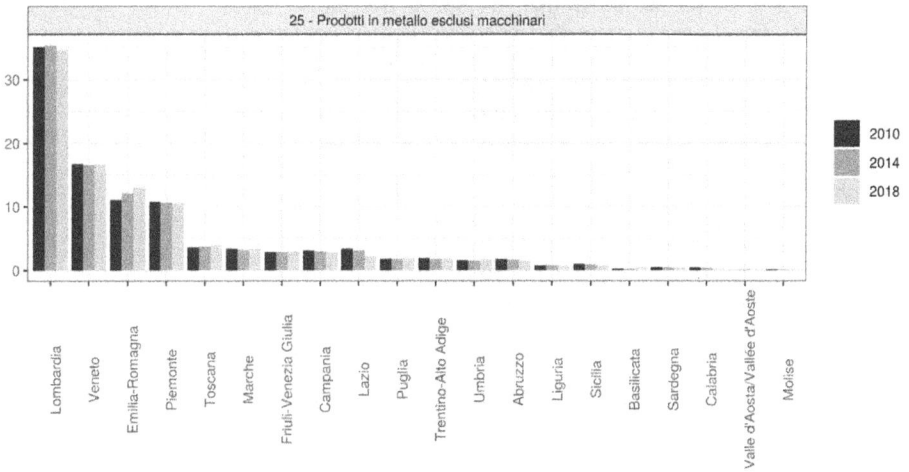
25 - Prodotti in metallo esclusi macchinari

How does Industry 4.0 affect the relationship between centre and periphery? The case of manufacturing industry in Germany

Samuel Greef ⓘ and Wolfgang Schroeder

ABSTRACT
In Germany, the debate on economic challenges and strategic orientation is strongly focussed on the industrial sector as the backbone and anchor of the German economy. In relation to the digitalization of the manufacturing industry, the term Industry 4.0 is used. The concept focuses primarily on possibilities for optimizing processes of production and product innovation. It thus aims at incremental rather than disruptive developments. Incremental digital development that enables synergies between existing regional strengths and the potentials of Industry 4.0 bears the danger of increasing rather than reducing regional disparities. The expansion of Industry 4.0 can especially be found in strong industrial centres. Many implementation examples are located in the vicinity of university towns, in regions with high population density as well as high expenditure on research and development by large industrial companies. As a result, there are hardly any shifts between the existing industrial centres and the periphery with its weak industrial base. So far, Industry 4.0 seems to have contributed little to reducing existing regional disparities.

Introduction: the varying significance of the industrial sector for German regions

Today's debate about where the economy is headed in the future is determined by the buzzword digitalization. The radical changes associated with digitalization are also reflected in the socio-political debates that focus on the relationship between economic-technical digitalization and its social effects in various areas. In the international context, the debate on progressive digitalization is focussed primarily on the service and knowledge sectors. It is, therefore, consistent with the plethora of narratives about the post-industrial service society, which dominates the discourse in the OECD countries. In contrast, Germany is pursuing its own path with a clear focus on the issue of the digitalization of the industrial sector.

Not only the question of implementation and the status quo plays a decisive role in this context but also heterogeneity and disparities because the overall national situation is shaped on a multi-regional basis. Regional differences and unequal developments, especially between cities and rural areas, present a form of social problems and political challenges. Concerns about the growing disparity of society, increasing inequality, dwindling social cohesion and political polarization are also fuelled by economic structural change and uncertainties regarding the effects of digitalization. Germany is characterized by a disparate geographic distribution of industrial centres, especially with regard to high-quality production. For example, the final manufacturers in the automotive industry are concentrated in the southern German states (Baden-Württemberg and Bavaria), while the suppliers are located in the eastern states (Saxony and Thuringia) and in North-Rhine-Westphalia. The regional disparities, especially between urban, metropolitan and rural areas, raises the question of whether the efforts to digitalize the industry are more likely to amplify existing disparities, or whether Industry 4.0 provides new possibilities for peripheral zones to develop their industrial production. For this reason, the regional perspective on the digitalization is emphasized by means of a comparative analysis of the federal states of Baden-Württemberg, North-Rhine-Westphalia and Brandenburg.

When referring to the digitalization of the industry, the term Industry 4.0[1] is used. The debate on economic challenges and strategic orientation is therefore focussed on industry, due to the particular importance of the industrial sector 'as the backbone and anchor' for value creation, employment and the competitiveness of the export-oriented German economy (Schroeder 2017, 1). One characteristic of the first decades in the existence of the Federal Republic of Germany was the steady growth of the industrial sector. This situation changed in the 1970s and 1980s. Employment in the industrial sector initially stagnated and then slowly began to decline. At the same time, the demand increasingly shifted towards services. These shifts were part of an ongoing structural change that was accompanied by a discourse about the substitution of the industrial society by a service and knowledge society. However, this discourse obscured the continued relevance of the industrial sector, especially in Germany. Compared to other countries, not only in Europe, it is striking that over the past 25 years, the share of total added value created by industry in Germany has remained relatively stable at 25%. By contrast, in other traditional industrialized countries, a drastic reduction in industrial capacities has taken place over the last 25 years (Schroeder, Greef, and Schreiter 2017, 4). In the UK, the value added share of the industrial sector (including energy) fell from 21% in the early 1990s to just under 14% in 2018, in France from 20% to 13% and in the USA from 20% to 15% (OECD 2020).

The continuing importance of the industrial sector for the stability and growth of the German economy in the last decades is reflected in the development of employment. Between 2007 and 2019, the number of people employed in the manufacturing sector rose from 7.8 to almost 8.4 million – although the service sector grew faster in the same period (from 29.5 to 33.7 million) (Statistisches Bundesamt (Federal Statistical Office) 2020a). But the growth of non-industrial services is often the result of corporate outsourcing strategies. Moreover, the demand for industrial goods usually resonates directly with the service sector, where it generates orders and jobs. In 2018, more than half of the total service production was business-related (iW 2019). Of the 33.4 million people employed in the service sector, a total of 6.2 million were working in the narrow field of business-related services (Statistisches Bundesamt 2020b). Nevertheless, the importance of the industrial sector varies greatly at the regional level, just as there are considerable differences in socio-economic structures and thus ultimately also in innovation potential. Therefore, it makes sense to take a differentiated look at the stability of industrial production in Germany and how different the regional situations are. In 2018, the proportion of employees in the manufacturing sector in the German federal states (excluding the city states)[2] varied between 34.7% in Baden-Württemberg and 20.9% in Schleswig-Holstein. Within Baden-Württemberg, in 2016 the differences between the counties ranged from 52% in Tuttlingen to 19.9% in Tübingen (Statistische Ämter des Bundes und der Länder Länder (Federal and State Statistical Offices) 2019, 2020). An industry 4.0 strategy sensitive to regional differences should take these unequal starting conditions into account.

Industry 4.0 as a federal political strategy based on existing structures in the regions

The term Industry 4.0 has been used in Germany since 2011 to describe a cooperative digitalization and industrial strategy. The central goal is to further develop the industry and maintain and expand its competitiveness as the backbone of the German economy. The German Industry 4.0 strategy not only represents an alternative to the market-based Silicon Valley US model but also to the Chinese state-centered top-down control strategy. It is obvious that the German Industry 4.0 debate emerged from a defensive position. This can be seen in the German economy's structural weakness in competitiveness in the ICT sector compared to American and leading Asian competitors.

The strategy focuses on the traditional strength of the German economy: the industry with its engineering know–how and diversified quality production, which is regarded as the key to the strategic debates on the future of the economy. Therefore, the aim is to strengthen competitiveness in this sector in particular. The Industry 4.0 strategy has two central elements: firstly, a better interlinkage of the economic, technical, educational and labour market-oriented policy areas. Secondly, the state, trade unions, companies and the scientific community need to cooperate more closely. The revitalization of corporatist arrangements offers vital concepts and frameworks for linking technological and social innovations. At the same time, they can provide the basis for broad social acceptance and legitimation so that these goals can be tackled more dynamically. The aim is not only to facilitate a better coordination of the necessary activities but also to overcome the fixation of the digitalization debate on technology and economy. The issue of qualifications is of particular importance in this context. The possible loss of jobs and the simultaneous emergence of new occupational profiles (such as that of a data analyst) creates a considerable demand for skilled workers, qualifications and further training. It is the areas of cyber-physical systems, additive processes, robotics and wearables, that are considered by a majority to be central to both initial training and continuing vocational training (Pfeiffer et al. 2016, 123ff.).

In contrast to the Anglo-Saxon debate on a new production model, the Industry 4.0 strategy does not rely on the thesis of a disruptive revolution.[3] The changes brought about by digitalization are not a sudden rupture but an evolutionary process that takes place incrementally and can thus be shaped.[4] The differences are politically relevant: While representatives of the disruptive perspective emphasize the creation of fundamentally new production and business models, the German discourse on Industry 4.0 focuses on possibilities for optimizing processes of production and product innovation that have been practiced for some time – depending on the industrial specialization. Therefore, digital development that enables synergies between existing regional strengths and the potentials of Industry 4.0 bears the danger of increasing regional disparities.

This is particularly the case if the digitalization of the industrial sector only contributes to a strengthening of the existing industrial centres instead of creating new opportunities for less industrialized regions to catch up. The answer to this question depends decisively on the design of the Industry 4.0 policy. A strategy that seeks to catch up development in regions with weak industries would have to be long-term in nature and initiate appropriate investments, because the decisive factor is how well the regions and their structures are prepared for the application of Industry 4.0.

Regional disparities and divergent opportunities: centre–periphery problems in manufacturing industries

Since the Hanover Fair 2011, the digitalization efforts labelled Industry 4.0 have made steady progress. 'Industry 4.0 as a concept has (…) become a real hype in the manufacturing industry' (Capgemini 2017, 4). This can also be seen in an annual survey which examines the digital transformation of the German industry since 2014 and summarizes the results in the Industry 4.0 Index, covering approximately 450 companies, of which two thirds can be assigned to the industrial sector. Since 2014, this index has risen from 16 to 42 points, although it only increased by one point from 2017 to 2018 (Staufen 2018, 7). This makes it clear that the hype surrounding Industry 4.0 alone is not enough. The underlying concept requires a continuous strategy. 'The digital transformation is not a quick fix, but a long-term commitment and a strategic necessity' (Capgemini 2017, 32). The stagnating Index is reflected in the faltering implementation of Industry 4.0 goals. Between 2014 and 2018, the share of companies pursuing 'single operational projects regarding Industry 4.0' rose from 15% to 43%, but even in 2018 only 8% of the companies surveyed implemented Industry 4.0 comprehensively. Over the same period, however, the share of companies that have 'not yet specifically' engaged with Smart Factory and Industry 4.0 decreased from 34% to 9%. Of the companies surveyed, the highest percentage of comprehensive operational implementation is in the automotive industry, at 18%, however, including single operational projects, with 64% the electrical industry is ahead of the automotive industry with 48% and the mechanical engineering and plant construction industry with 56% (Capgemini 2017, 12). The main centres for the implementation of Industry 4.0 across all sectors are BaWü (especially the area around Stuttgart), North-Rhine-Westphalia with the Rhine-Ruhr metropolitan region and Bavaria (the Munich metropolitan region in particular) (Plattform Industrie 4.0 n.d.).

The slow implementation of Industry 4.0 projects in all industrial sectors coincides with considerable regional disparities on two levels. The first level concerns the significance of the respective sub-sectors. When comparing profits in the automotive, mechanical engineering, electronics and optics sectors at the level of the federal states, clearly regional disparities become visible when revenues in three Eastern German federal states fall even behind those of the city states (Statistisches Bundesamt 2020c) (see Figure 1).

Nevertheless, digitalization acts as a driver for an increasing interlinkage of value creation processes across industry and sector boundaries. As a result, the differences

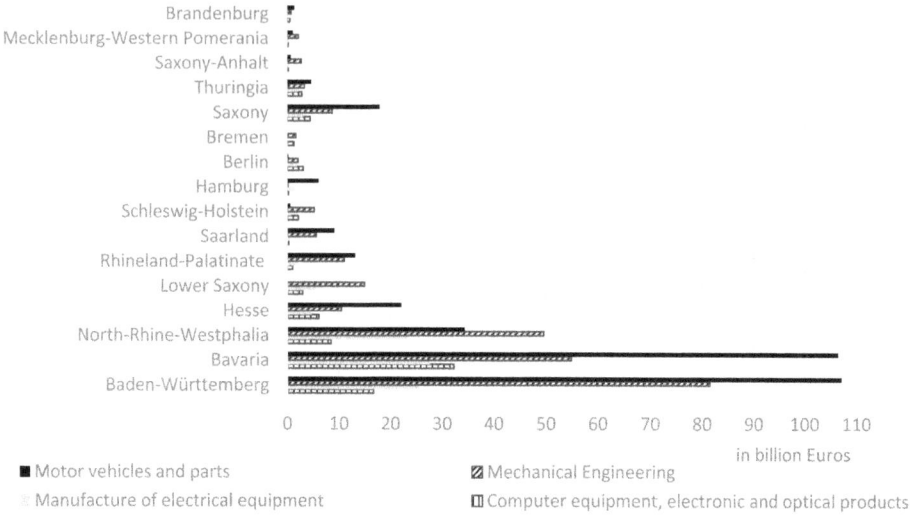

Figure 1. Revenue in manufacturing industry 2018. Source: Statistisches Bundesamt (2020); authors' presentation.

between individual sectors and between services and industry are becoming less relevant (Bertschek 2015).

The second level concerns the current state of digitalization of the 'hidden champions' – the small and medium-sized enterprises (SMEs) – which are particularly important for Germany. A study by IW Consult for 2017 concludes that the starting position varies considerably from region to region: The Digital Index based on externally observable indicators shows a 'very clear West–East divide and an urban-rural divide' (Lichtblau et al. 2018, 19). While large cities (like Frankfurt, Hanover, Stuttgart or Munich) were 23 points above the average index value of 100, low-density rural counties were 14 points below. Of 47 rural counties in Germany, there are 35 (74%) in Eastern Germany fell into the lowest quantile of the Digital Index, while none fell into the highest. All 10 regions in Eastern Germany in the highest quantile were non-county cities[5] (Saxony: Leipzig, Dresden; Saxony-Anhalt: Magdeburg; Thuringia: Eisenach, Erfurt, Weimar, Jena; Brandenburg: Potsdam; Mecklenburg-Western Pomerania: Schwerin, Rostock). By contrast, only 22.5% (55) of the 247 counties in Western Germany were in the lowest quantile. The starting positions for the implementation of an Industry 4.0 strategy thus differ considerably from region to region.

With digitalization, the importance of spatial distance decreases in many cases. However, such a correlation may not apply to the manufacturing industry. In fact, even the opposite effect is possible: if with the ongoing digitalization of industry, additional tasks from the low and medium qualification segment are automated by more sensitive autonomous robots and AI, this could mean the elimination of those local jobs in structurally weak regions. At the same time, highly qualified job profiles are becoming more important, which can theoretically be carried out from any place, and thus from locations that are more attractive for employees. The last point is expressed in a similar manner in the National Industrial Strategy 2030: 'However, in view of the disruptive nature of many changes, the risk exists that new, innovative and future-viable jobs will not necessarily be created in (…) regions in which existing jobs are lost as a result of technological progress and increases in productivity' (BMWi 2019, 7). There is a risk that rural regions with a high proportion of employees in manufacturing will be especially affected by job losses, whereas metropolitan regions with a good digital infrastructure and knowledge-intensive R&D sectors will benefit most.

Regional disparities: Industry 4.0 in Baden-Württemberg, North-Rhine-Westphalia and Brandenburg

Economic conditions, regions and locations are important for the digitalization of industry, and therefore also for the implementation of the concepts behind Industry 4.0, for two reasons. Firstly, digital modernization and the introduction of completely new networked production methods does not only depend on the perceived necessity and thus the willingness to implement them but requires extensive financial resources – consequently the ability to make targeted investments. Secondly, Industry 4.0 does not start in a vacuum. Without existing products and production structures, a digitalization strategy can be based on implementation costs and planning horizons increase. The case selection takes this into account. The starting position of the compared Federal States, Baden-Württemberg (BaWü), North-Rhine-Westphalia (NRW) and Brandenburg, differ considerably regarding not only industrial structures, research and innovation potential as well as the structure of the workforce and population density. With 1.5 million employees, predominantly in the automotive and mechanical engineering sectors as well as in electrical engineering, BaWü is the largest industrial location in Germany. These three sectors also play a major role in the consistently high share of gross value added in industry and the importance of the export sector. 'The state is proud of its decentralized and mainly medium-sized economic structure' (Wehling 2013, 342).

In NRW, the industry structure has diversified: while coal and steel (in the Ruhr area) and the textile industry (Rhineland) have become less important, the production of investment goods, the chemical industry and vehicle construction have become more important. As the 'home of numerous industrial companies of worldwide reputation (…), NRW is also increasingly becoming a state of the medium-sized sector' (Andersen and Woyke 2013, 399).

The starting position of Brandenburg for the implementation of Industry 4.0 concepts is completely different from that of NRW and BaWü. Brandenburg is rural and thinly populated. The 'industrial profile with lignite mining, iron and steel production, petrochemicals, metallurgy, electrical engineering/electronics, optical industry, and mechanical and vehicle engineering' developed in times of the German Democratic Republic was shattered during the de-industrialization following German reunification' (Franzke 2019, 4). Today, many innovative SMEs play a central role, whereas only a few major enterprises are still relevant.

In terms of surface area, the Brandenburg counties are considerably larger and less densely populated than those in NRW and BaWü. Between the latter states, the districts are similarly distributed in terms of surface area but are even more densely populated in NRW (see Figure 2).

The GDP of the 3 federal states in 2017 was 45,064 euros per inhabitant in BaWü (113.7% compared to the German average), 38,276 euros in NRW (96.5%) and 28,473

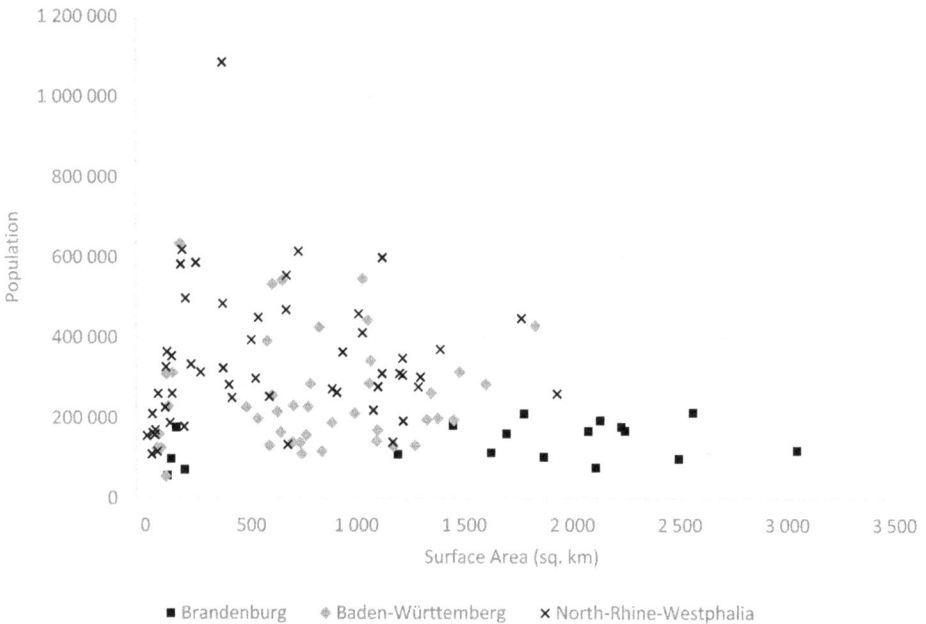

Figure 2. County population density 2018. Source: Statistische Ämter des Bundes und der Länder (2019); authors' presentation.

euros in Brandenburg (71.8%). These differences are relevant because strong federal states can and will invest more. Economically strong states are better prepared to support their industries to match the coming challenges. This is reflected in the volume of the federal states' investments programmes for digitalization. NRW is investing 640 million euros solely in the digital economy until 2020. Brandenburg by comparison invests 450 million euros in 200 digitalization projects, of which only a quarter is related to the economy. They focus on SMEs in nine regional clusters that are already of particular economic importance, five of them in the immediate vicinity of Berlin. In NRW, the focus is on the connecting industry, SMEs and start-ups through six regional digital hubs. These are explicitly oriented towards existing strong industrial bases in order to bring established companies together with the ideas of young, creative start-ups. In BaWü, 10 regional digital hubs are being funded with 10 million euros, which are also intended to especially support SMEs in their digitalization efforts. In addition, 25 million euros were provided for digital start-ups. Both are part of a total volume of 323.3 million euros that BaWü invested from 2017 to 2019. Intelligent mobility, including autonomous driving and battery cell production, accounted for 15% of this amount. 14% was invested in Economy 4.0 and 5G as the basis for digital production environments. This included, for example, digitalization bonuses for small companies that have implemented specific digital transformation projects. The manufacturing sector accounted for 26% of the approved projects. In addition, 15 digital transfer projects were financed, of which 40% were located in the industrial sector.

Therefore, the potential and need for industrial modernization is linked to the importance of sectors with an affinity for Industry 4.0, which is highest for manufacturers. The share of gross value added accounted for by the manufacturing sector differs significantly in one of the three states. In Brandenburg (27%) and NRW (28%), the share in 2018 was below the Germany-wide average of 31%. BaWü, on the other hand, had a share of 40% of gross value added in the manufacturing sector. The situation was somewhat different for the investment ratios in 2016.[6] They were 23.6% in Brandenburg, 21.5% in BaWü and only 16.1% in NRW (average in Germany: 20.3%). As a result of the considerable differences in the GDP level, NRW leads the three federal states in terms of gross investments with 106.6 billion euros (only outdistanced by Bavaria), ahead of BaWü with 102 billion euros and Brandenburg with only 16.2 billion euros. Gross investments in the manufacturing industry have increased by 17% from 2015 to 2018 (from 294.4 to 345.4 billion euros) (cf. Statistisches Bundesamt 2019a, 18ff., 32, 65). In terms of the volume of investment funds alone, the modernization potential, and thus the implementation probability of Industry 4.0, appears to be significantly higher in NRW and BaWü than in Brandenburg. Weak regions lag behind and do not have the same capabilities to prepare for an innovative future.

This is also reflected in the figures for research and development (R&D) and patents. R&D is not only a central part of the value chain. They define the cornerstones for a consistent implementation of Industry 4.0. In BaWü particularly the administrative region of Stuttgart stands out in the area of R&D. With an R&D expenditure of almost 15 billion euros in 2017 and almost 80,000 employees, this region alone is far ahead of the whole NRW region with nearly 60,000 employees and 8.5 billion euros expenditure. The data for NRW (3 billion euros and 17,600 employees) roughly correspond to the three other administrative districts in BaWü taken together: Karlsruhe with almost 4 billion euros and 25,000 employees, Tübingen with 3 billion euros and 17,000 employees and Freiburg with 1.5 billion euros and 11,000 employees. In NRW administrative district of Düsseldorf that lay ahead of the district of Cologne with 2.4 billion euros and 17,500 employees. They were followed by Arnsberg and Detmold with both 1.2 billion euros and about 10,000 employees. The administrative district of Münster was weakly represented regarding application examples for Industry 4.0 and came last with just under 540 million euros and 4,000 employees; however, it was still ahead of Brandenburg in both areas (Stifterverband für die Deutsche Wissenschaft 2019, 51). In the south of Brandenburg, the resources spent[7] on R&D in 2017, at 325 million euros and just under 3500 employees, were significantly higher than the barely 80 million euros and less than 900 employees in North Brandenburg. All of Brandenburg is still behind the weakest administrative district in NRW (Münster with 536 million euros and 3800 employees) and far behind the weakest in BaWü (Freiburg with 1.5 billion euros and 11,000 employees). Brandenburg, with a total of 88.3 patents per 1 million inhabitants filed with the European Patent Office in 2012, was also far behind the 148.4 (Münster) to 237.3 (Detmold) patents in NRW or the 359.6 (Tübingen) to 480.7 (Stuttgart) patents in BaWü (eurostat 2020). Do these different starting positions affect the implementation of Industry 4.0 concepts?

Implementation of Industry 4.0

An exemplary illustration of the implementation of Industry 4.0 is provided by the Platform Industry 4.0 (Plattform Industrie 4.0 n.d.). With regard to the manufacturing industry, a comparison between the federal states shows that no other state registers as many application examples as BaWü: 58, of a total of 181 examples can be found here. NRW follows in second place with 40 examples, while Brandenburg is in sixth place with only five examples (the same as Berlin, Saxony and Saarland). The examples are distributed among the different company size categories in Table 1

Despite the differences in the number of Industry 4.0 examples, it is evident that SMEs play a role in all three federal states, but the numbers do not reflect their overall importance for the economy. Industry 4.0 has so far remained mainly a phenomenon in large-scale industry.

The distribution of the regional locations where Industry 4.0 has been applied also shows a pattern: With the exception of a concentration in the NRW-Ruhr area between Düsseldorf and Dortmund (14), they are not located in the regions where the industrial density, measured by the share of employees in the manufacturing industry, is the highest. Instead, there are concentrations around the university cities. Stuttgart (8), Karlsruhe (5) and Mannheim/Heidelberg (6) in BaWü and Bielefeld/Paderborn (11), Aachen (4) and Bonn (5) in NRW. In Brandenburg, four of the five application examples are located in the industrial belt in Berlin's immediate vicinity. 'The concentration of economic activities in Germany's major cities and their surrounding areas has thus increased at the expense of peripheral rural areas in particular' (Küpper and Peters 2019, XIX). An Industry 4.0-Strategy that aims primarily at already strong industrial centres may accelerate this process of divergence. The current examples of the application of Industry 4.0 seem to confirm this. They speak against a reduction of disparities between densely populated urban centres and rural peripheral areas. The beacons of Industry 4.0 shown on the map are not to be found in the less densely populated rural areas but can be located in the postcode areas with a high population density (see Figure 3).

A deeper insight into the use of new technologies subsumed under the term Industry 4.0 is provided by the Transformationsatlas[8] of IG Metall.[9] The nationwide[10] survey asked about the use of seven central physical (collaborating robots, 3-D printing and data glasses/tablets) and digital technologies (digitally communicating production means, automated production planning/management, AI for activity automation and IT/AI for remote diagnosis/maintenance of machines). This shows that the areas of digitally communicating production means and automated production planning and management are an issue for about 70% of all companies, followed by data glasses/tablets and remote diagnosis/maintenance for more than 60%. In contrast, AI and collaborative robots play a role for only one third of all companies and 3D-printing for just under half. A second question focuses on the current stage and implementation of digitalization. The area of product digitalization is the most advanced, with over 60% of companies stating

Table 1. Application examples Industry 4.0.

Company size category (no. of employees)	Number of applications of I4.0 in the manufacturing industry		
	BaWü	NRW	Brandenburg
1–250	21	15	2
250–5000	18	7	2
5000–15,000	15	11	1
15,000+	5	7	0

Source: Plattform Industrie 4.0 (n.d.) (as per 23 January 2020).

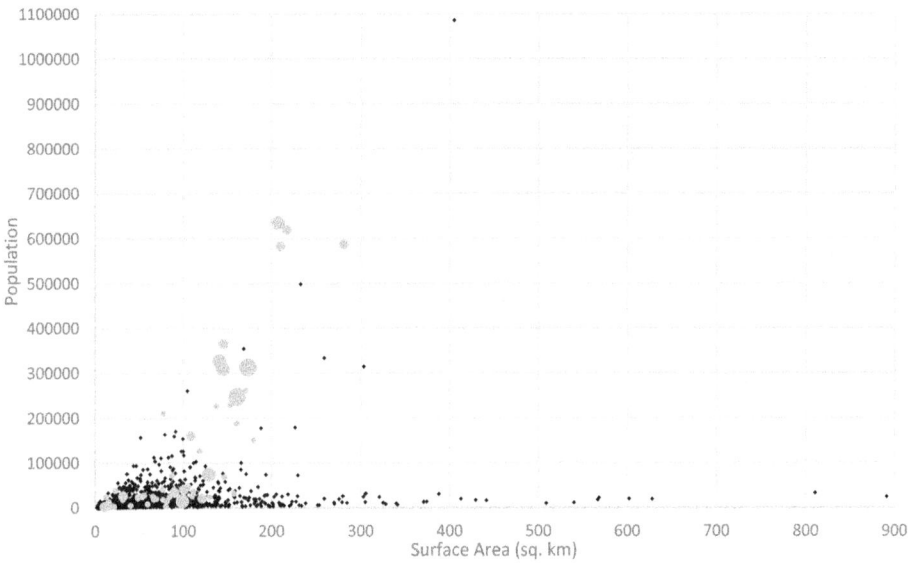

Figure 3. Locations of Industry 4.0 application examples in the manufacturing industry by postcode areas. Source: Statistische Ämter des Bundes und der Länder (2019); authors' representation. Note: black-dots: a total of 1625 postcode areas; grey-dots: depending on size: 1–4 Industry 4.0 examples.

that digitalization has been implemented in parts or to a large extent. The other areas are the digitalization of manufacturing processes (54%), maintenance processes (48%) and administration processes (57%).

The Transformationsatlas shows distinct differences between the federal states. BaWü lies above the German average in the application of new technologies (+4% automated production, +5% AI automation, +11% 3D-printing). At the same time, however, the stage of implementation of digitalization is below average (−5% products and maintenance processes and −8% manufacturing and administration processes). In contrast, NRW is clearly above the national average in all areas when it comes to the implementation of digitalization (+3% products and maintenance processes, +4% administration processes and +6% manufacturing processes). NRW thus plays a pioneering role on the one hand: in incremental process digitalization. On the other hand, the state is also lagging slightly behind in the introduction of new innovative concepts and possibilities. At the same time, NRW lies above the German average in the application of new technologies only in the areas of digitally communicating production means (+3%), as well as slightly in AI and remote diagnosis/maintenance (+0.6%). Brandenburg legs behind in four of the seven areas (−5% collaborating robots and AI automation, −7% data glasses/tablets, −9% 3D-printing) and is only above-average in digitally communicating production means (+3.7%). At the same time, it only achieves the national average in the areas of production and administration digitalization (−3% maintenance processes and −5% manufacturing processes) – however, in comparison to NRW, this result is significantly better. These results can be explained by the greater regional disparities within Brandenburg. Few large companies as pioneers of digitalization in Berlin's affluent suburbs are contrasted by a large number of small companies and crafts enterprises in rural areas that are lagging behind digital development or see it as irrelevant for their business.

Despite the differences, the percentage distances are all in all relatively small. The reasons for this can be derived from the Smart Factory research project (2016). If, for example, 'Industry 4.0 is nothing new, but rather the continuation, so to speak, of the technological development we have known in the past' (Kohler and Gscheidle 2019, 169). The operational reality often consists of incremental digitalization. However, a 'continuous modernization of production plants' is nowhere near the overall concept of Industry 4.0 – and therefore the 'smart factory – which controls production itself is more of a vision than reality' (Kohler and Gscheidle 2019, 173). Consequently, in order to enable a digital transformation of companies in the sense of the Industry 4.0 concept, two essential competencies are required: 'on the one hand, to develop digital skills, processes, products and business models digitally, and on the other hand the transformational skills to initiate appropriate change processes in the company' (Obermaier 2019, 8). This requires corresponding strategic management and a supportive organizational culture in industrial companies. By no means all established industrial companies have yet understood that 'in a digital world, they must act more like a software company, that they need more start-up culture and to make considerable investments in platforms, ecosystems and new business models' (Fend and Hofmann 2018, 298).

Industries between 3.0 and 4.0: patterns, explanations and perspectives

What general conclusions can be drawn from the above analysis of the three federal states? At first glance, the comparison between Brandenburg on the one hand and BaWü and NRW on the other seems to correspond to a logic of differentiation between East Germany and West Germany. In this respect, the federal states examined differ in their starting position both on a quantitative level (number of large companies) and on a qualitative level (economic success of the companies). However, the analysis above has shown that, in terms of innovation potential, the application of digital technologies and the implementation of Industry 4.0 concepts, there is no primarily East German logic. The potentials for Industry 4.0 rather vary between regional disparate starting positions. These are mainly determined by (non-)existent industrial structures and thus follow a logic of differences between urban centres and rural periphery. Therefore, region-specific settings play the main role, in which structurally strong industrial locations with high expenditure on research and development lead the way towards Industry 4.0. 'Regions without equivalent research structures and industrial innovation potential are coming under increasing pressure in their developments' (Jakszentis and Hilpert 2007, 28). Due to their poorer starting position, structurally weak rural regions in both East and West Germany have a lower technological-industrial innovation potential. The only difference is that this type of region is more common in Eastern Germany than in many Western German states.

Furthermore, the size of companies plays a major role. The issue of digitalization has arrived in the majority of large companies – even examples of the application and implementation of process digitalization and new digital products can be found. However, this perspective obscures the importance of SMEs for the German economy. In 2017 they accounted for 99.3% of all companies, employed 60.8% of all workers, made 43% of all gross investments, and generated 33.2% of total revenue (Statistisches Bundesamt 2019b). This also applies, albeit at a lower level, to the manufacturing sector (42.9% of employees and 30.6% of revenue in 2016 were accounted for by SMEs) (Lichtblau et al. 2018, 7). In relation to the implementation of Industry 4.0, these 'hidden champions' are of great interest. For a long time, SMEs, in particular, were said not to pay sufficient attention to the issue of digitalization. Even in 2016, the proportion of SMEs that did not make any or only a limited effort to incorporate digitalization into their corporate strategy was still over 45% – compared to 26% among large companies (Lichtblau et al. 2018, 14). Only in the last two years, a reorientation can be observed. Nevertheless, the proportion of SMEs who consider Industry 4.0 to be of (very) high strategic importance only rose from 37% to 49% between 2015 and 2018. Especially successful medium-sized companies attach great importance to Industry 4.0. 65% of SMEs stated that Industry 4.0 was of strategic importance (Becker, Ulrich, and Botzkowski 2019, 99f.). At the same time, however, only 37% considered Industry 4.0 to be fairly or very important for their company. Yet, 66% assume that the importance will increase in the future. The difference between SMEs and large companies in their assessment of the situation is also reflected in the actual application of digital technologies: it increases with company size. For many indicators (such as the use of cloud computing, big data, robotics/sensor technology or supply chain management), therefore 'the intensity of use is in part significantly higher in larger companies than in SMEs' (Lichtblau et al. 2018, 21).

Although there are isolated beacons for the implementation of Industry 4.0, it does not seem as if it will be implemented across the board in the medium term. The current state of implementation of Industry 4.0, therefore, counters the theory of a technological disruption that is taking place or has taken place. The development is characterized less by a revolutionary rupture than by an evolutionary implementation through incremental modernization and gradual migration to new technologies (Lichtblau et al. 2018, 98f.). In the versioning, this could be referred to as Industry 3.5: the smart factory is still a vision of the future. Thus, existing industrial structures and not an East German logic widely determine the application of Industry 4.0.

The lack of an East German logic is also evident regarding the possible effects of digitalization on employment and the labour market. The debate around the danger of job loss due to digitalization, started with a study by Frey and Osborne (2013), has also affected Germany and the discussion about the vision of Industry 4.0. All IAB[11] country studies on the substitutability potential of employees show a distinct mutual correlation for the three federal states examined here: as the proportion of employees in the manufacturing sector increases, so does the impact of a high substitutability potential through digitalization. This correlation is not only found at the federal state level but also clearly within the counties. The degree to which the high substitutability potential affects the counties in Baden-Württemberg rises from 10% in counties with a manufacturing share of 10% of all employees to more than 30% in counties with an employment share of almost 60%. In Berlin-Brandenburg, the same effects are evident. The correlation can be explained, among other things, by the fact that professions in manufacturing and production engineering are especially at risk (Hafenrichter et al. 2016, 18, 26; Seibert, Jost, and Wiethölter 2019, 9, 13, 16; Sieglen 2018, 14). In addition, the substitutability potential has increased at an above-average rate, particularly in the business-related service occupations (see Wydra-Somaggio 2019, 7f.). 'The regional share of employees with high substitutability potential depends largely on the specific industry and occupational structure in a federal state' (Wydra-Somaggio 2019, 13).

This perspective clearly shows once again that the issue of qualification as well as initial and further vocational training in the context of Industry 4.0 plays a central role in order to meet the new requirements of a digitalized industry and thus enable a broader implementation of the concepts associated with Industry 4.0. It is no coincidence that the qualification of employees is most frequently (60%) also named by SMEs themselves as the challenge for medium-sized enterprises in the Industry 4.0 context (Becker, Ulrich, and Botzkowski 2019, 104). At the same time, however, a look at the substitutability potentials shows that qualification is also one of the central issues that needs to be addressed so that technological and social innovations as two sides of the same coin can be sustainably connected.

Conclusion

It is important to keep a sensitive eye on the development of regional disparities. So far, the digitalization of processes, products and services seems to have contributed little to reducing inequalities. Rather, an expansion of Industry 4.0 can be found especially in the strong centres of German industry. One of the factors is the continuing low affinity of SMEs for digitalization. Especially with regard to SMEs and in view of the provision of digital infrastructure in rural areas, increased efforts are required in order to actually exploit the potentials of digitalization for a reduction of regional disparities. The second factor is that the majority of the industry's digitalization is progressing incrementally, following path-dependent established modernization patterns. As a result, there are hardly any shifts between the existing industrial centres and the periphery with its weak industrial base.

With regard to the regions in Germany, the diverse production profiles coincide with the uneven geographical distribution and locations of Industry 4.0 application examples. Regions with a production profile essentially based on unskilled labour often lack large innovative, research-intensive companies. On the other hand, regions with such companies also have considerably more research and development departments. This is reflected in the locations of application examples insofar as many of these examples are located in the vicinity of university towns and in regions with high expenditure on research and development. Incremental innovations through Industry 4.0 implementation thus bears the risk of increasing rather than reducing existing regional disparities.

The economic structural change and the digitalization of the industry is indeed related to the question of regional disparities. However, we must not lose sight of the fact that the unequal living conditions and polarization between cities and rural areas have many underlying causes. In this respect, much can already be achieved if industrial policy contributes to preventing an exacerbation of existing inequalities, but it is hardly an appropriate instrument for reducing these disparities on its own.

Notes

1. The term Industry 4.0 was introduced at the Hannover Messe 2011. It originates from acatech, the German Academy of Science and Engineering. They proposed the future project Industry 4.0 as a course of action to enable Germany to 'maintain its position as a production location even in a high-wage region' in the global competition, which

means 'getting fit for the fourth industrial revolution driven by the Internet' (Kagermann, Lukas, and Wahlster 2011).

The version number 4.0 originates from software development and adopts a counting scheme used for the historical course of industrial development. It begins with the first industrial revolution (steam engine), followed by the second (assembly line) and the third (computer) (Schwab 2016, 16f.). The essence of the fourth revolution is the new quality of self-regulated cyber-physical systems (CPS), which refers to the ubiquitous, decentralized networking and autonomous communication of IT-controlled machines, products and people. Industry 4.0 aims at the 'process efficiency of industrial value creation' and at 'product innovations in the form of intelligent and interconnected products' (Obermaier 2019, 4f.).

2. Germany is divided into 16 federal states. Berlin, Hamburg and Bremen are called city states as their territory covers only a single city and the surrounding area.
3. Nevertheless, terms such as disruption or disruptive innovation can be found frequently in the publications of the Federal Government and ministries.
4. An evolutionary process can nevertheless be characterized by dynamic phases of development and can also have retrospectively revolutionary effects in its long-term impact (Obermaier 2019, 3).
5. 10,795 municipalities represent the lowest level of administrative structure. They are either part of one of the 294 counties or one of the 107 non-county cities. Non-county cities have enough inhabitants and administrative power to take over the tasks of a county municipality as well as the tasks of a county.
6. Share of gross investment in GDP in % (in respective prices).
7. Internal expenditures of companies for R&D.
8. In 2019, IG Metall surveyed nearly 2000 companies with 1.7 million employees about the stage of their digital transformation. The survey is called Transformationsatlas (Transformation Atlas).
9. The IG Metall (Industriegewerkschaft Metall) (Industrial Union of Metalworkers) is the largest trade union in Germany (with 2.26 million members in 2019). The union represents workers in the manufacturing and industrial production, in the engineering and the electrical sector as well as workers in the wood, plastics, textiles and clothing industry.
10. The IG Metall regions are not always congruent with the federal states, for example Brandenburg and Saxony form one region.
11. The statutory mandated IAB (Institut für Arbeitsmarkt- und Berufsfroschung) (Institute for Employment and Occupational Research) is the scientific research and consulting institution of the Federal Employment Agency.

Disclosure statement

No potential conflict of interest was reported by the author(s).

ORCID

Samuel Greef ⓘ http://orcid.org/0000-0002-1737-5541

References

Andersen, U., and W. Woyke. 2013. "Land Nordrhein-Westfalen." In *Handwörterbuch des politischen Systems der Bundesrepublik Deutschlan*, edited by U. Andersen and W. Woyke, 398–405. Wiesbaden: Springer.
Becker, W., P. Ulrich, and T. Botzkowski. 2019. "Industrie 4.0 im Mittelstand – Handlungspotenziale und Umsetzung." In *Handbuch Industrie 4.0 und Digitale*

Transformation. betriebswirtschaftliche,technische und rechtliche Herausforderungen, edited by R. Obermaier, 3–46. Wiesbaden: Springer.

Bertschek, I. 2015. *Industrie 4.0: Digitale Wirtschaft – Herausforderung und Chance für Unternehmen und Arbeitswelt*. München: ifo Institut.

BMWi. 2019. *Industrial Strategy 2030. Guidelines for a German and European industrial policy.* Berlin: BMWi. Accessed March 5, 2020. https://www.bmwi.de/Redaktion/EN/Publikationen/Industry/industrial-strategy-2030.pdf?__blob=publicationFile&v=7

Capgemini. 2017. *Industrie 4.0 – Eine Einschätzung von Capgemini Consulting. Der Blick über den Hype hinaus.* Berlin: Capgemini.

eurostat. 2020. Patent applications to the European patent office (EPO) by priority year by NUTS 2 regions, Eurostat. Accessed February 17, 2020. https://ec.europa.eu/eurostat/databrowser/view/tgs00040/default/table?lang=en

Fend, L., and J. Hofmann. 2018. *Digitalisierung in Industrie-, Handels- und Dienstleistungsunternehmen*. Wiesbaden: Springer Gabler.

Franzke, J. 2019. "Land Brandenburg." In *Handwörterbuch des politischen Systems der Bundesrepublik Deutschland*, edited by U. Andersen, J. Bogumil, S. Marschall, and W. Woyke, 1–8. Wiesbaden: Springer VS.

Frey, C., and M. Osborne. 2013. *The Future of Employment: How Susceptible Are Jobs to Computerisation?, Oxford Martin Programme on the Impacts of Future Technology.* Oxford: University of Oxford. Accessed December 3, 2019. https://linkinghub.elsevier.com/retrieve/pii/S0040162516302244

Hafenrichter, J., S. Hamann, O. Thoma, T. Buch, and K. Dengler. 2016. *Digitalisierung der Arbeitswelt. Folgen für den Arbeitsmarkt in Baden-Württemberg, IAB-Regional 3/2016.* IAB: Nürnberg.

iW. 2019. Anteil Dienstleistungen an der Bruttowertschöpfung. Accessed April 19, 2020. https://www.deutschlandinzahlen.de/?452

Jakszentis, A., and U. Hilpert. 2007. *Wie spezifisch sind die Entwicklungen in Ostdeutschland? Angleichung der industriellen Modernisierungsprozesse in Ost- und Westdeutschland. Am Beispiel von Jena und Göttingen, Rostock und Kiel, Chemnitz und Braunschweig.* Frankfurt: OBS.

Kagermann, H., W. Lukas, and W. Wahlster. 2011. Industrie 4.0: Mit dem Internet der Dinge auf dem Weg zur 4. industriellen Revolution – ingenieur.de, ingenieur.de. Accessed January 7, 2020. https://www.ingenieur.de/technik/fachbereiche/produktion/industrie-40-mit-internet-dinge-weg-4-industriellen-revolution/

Kohler, H., and K. Gscheidle. 2019. "Deutschland/Baden-Württemberg: Ergebnisse und Interpretation einer Expert_innen-Befragung zu den Veränderungen der Arbeit in den Unternehmen aufgrund von Digitalisierung und Technologisierung." In *Smart Factory und Digitalisierung: Perspektiven aus vier europäischen Ländern und Regionen*, edited by D. Bürkardt, H. Kohler, N. Kreuzkamp, and J. Schmid, 165–175. Baden-Baden: Nomos.

Küpper, P., and J. Peters. 2019. *Entwicklung regionaler Disparitäten hinsichtlich Wirtschaftskraft, sozialer Lage sowie Daseinsvorsorge und Infrastruktur in Deutschland und seinen ländlichen Räumen, Thünen Report 66.* Braunschweig: Thünen-Institut. Accessed December 22, 2019. http://d-nb.info/1177156938/

Lichtblau, K., T. Schleiermacher, H. Goecke, and P. Schützdeller. 2018. Digitalisierung der KMU in Deutschland, IW Consult.

Obermaier, R. 2019. "Industrie 4.0 und Digitale Transformation als unternehmerische Gestaltungsaufgabe." In *Handbuch Industrie 4.0 und Digitale Transformation. betriebswirtschaftliche, technische und rechtliche Herausforderungen*, edited by R. Obermaier, 3–46. Wiesbaden: Springer Gabler.

OECD. 2020. "Gross National Income: Value Added by Activity." OECD Data. Accessed July 22, 2020. http://data.oecd.org/natincome/value-added-by-activity.htm

Pfeiffer, S., H. Lee, C. Zirnig, and A. Suphan. 2016. *Industrie 4.0 - Qualifizierung 2025, Bildung.* Frankfurt: VDMA.

Plattform Industrie 4.0. n.d. *Landkarte Industrie 4.0.* Accessed June 24, 2019a. https://www.plattform-i40.de

Schroeder, W. 2017. *Industrie 4.0 und der rheinische kooperative Kapitalismus, WISO Direkt 03.* Bonn: FES.

Schroeder, W., S. Greef, and B. Schreiter. 2017. *Shaping Digitalisation. Industry 4.0 – Work 4.0 – Regulation of the Platform Economy, International Policy Analysis.* Berlin: FES.

Schwab, K. 2016. *Die Vierte Industrielle Revolution.* München: Pantheon.

Seibert, H., O. Jost, and D. Wiethölter. 2019. *Mögliche Auswirkungen der Digitalisierung in Berlin und Brandenburg, IAB-Regional 2/2019.* IAB: Nürnberg.

Sieglen, G. 2018. *Digitalisierung in Nordrhein-Westfalen: Substituierbarkeitspotenziale der Berufe 2016. Aktuelle Ergebnisse auf Basis einer Neubewertung der Substituierbarkeit von beruflichen Kerntätigkeiten, IAB-Regional 1/2018.* IAB: Nürnberg.

Statistische Ämter des Bundes und der Länder. 2019. *Erwerbstätige nach Wirtschaftsbereichen.* Accessed February 14, 2020. https://www.statistikportal.de/de/erwerbstaetige-nach-wirtschaftsbereichen

Statistische Ämter des Bundes und der Länder. 2020. *Arbeitnehmer nach.* Accessed February 14, 2020. https://www.regionalstatistik.de/genesis/online/data;sid=87497DEA26B732726B63E3895736208A.reg1?operation=abruftabelleBearbeiten&levelindex=2&levelid=1581696232975&auswahloperation=abruftabelleAuspraegungAuswaehlen&auswahlverzeichnis=ordnungsstruktur&auswahlziel=werteabruf&selectionname=13312-02-03-4&auswahltext=&werteabruf=Werteabruf

Statistisches Bundesamt. 2019a. *Volkswirtschaftliche Gesamtrechnung. Arbeitsunterlage Investitionen. 3. Vierteljahr 2019* Accessed February 15, 2020. https://www.destatis.de/DE/Themen/Wirtschaft/Volkswirtschaftliche-Gesamtrechnungen-Inlandsprodukt/Publikationen/Downloads-Inlandsprodukt/investitionen-pdf-5811108.pdf?__blob=publicationFile

Statistisches Bundesamt. 2019b. *Kleine und mittlere Unternehmen.* Accessed February 16, 2020. https://www.destatis.de/DE/Themen/Branchen-Unternehmen/Unternehmen/Kleine-Unternehmen-Mittlere-Unternehmen/Tabellen/wirtschaftsabschnitte-insgesamt.html. Library Catalog: www.destatis.de

Statistisches Bundesamt. 2020a. *Erwerbstätige und Arbeitnehmer nach Wirtschaftsbereichen.* Accessed February 14, 2020. https://www.destatis.de/DE/Themen/Arbeit/Arbeitsmarkt/Erwerbstaetigkeit/Tabellen/arbeitnehmer-wirtschaftsbereiche.html

Statistisches Bundesamt. 2020b. *Erwerbstätige im Inland nach Wirtschaftsbereichen.* Accessed February 14, 2020. https://www.destatis.de/DE/Themen/Wirtschaft/Konjunkturindikatoren/Volkswirtschaftliche-Gesamtrechnungen/vgr010.html

Statistisches Bundesamt. 2020c. *Beschäftigte und Umsatz der Betriebe im Verarbeitenden Gewerbe.* Accessed February 15, 2020. https://www-genesis.destatis.de/genesis/online/data?operation=abruftabelleBearbeiten&levelindex=0&levelid=1581780483133&auswahloperation=abruftabelleAuspraegungAuswaehlen&auswahlverzeichnis=ordnungsstruktur&auswahlziel=werteabruf&code=42271-0011&auswahltext=&werteabruf=Werteabruf

Staufen. 2018. *Industrie 4.0. Deutscher Industrie 4.0 Index 2018.* Köngen: Staufen.

Stifterverband für die Deutsche Wissenschaft. 2019. *Forschung und Entwicklung in der Wirtschaft. Zahlenwerk 2019.* Essen: SV Wissenschaftsstatistik.

Wehling, H. 2013. "Land Baden-Württemberg." In *Handwörterbuch des politischen Systems der Bundesrepublik Deutschland,* edited by U. Andersen and W. Woyke, 341–347. Wiesbaden: Springer VS.

Wydra-Somaggio, G. 2019. *Warum die Digitalisierung manche Bundesländer stärker betrifft als andere.* Accessed December 22, 2019. https://www.iab-forum.de/warum-die-digitalisierung-manche-bundeslaender-staerker-betrifft-als-andere

Glowing cities and the future of manufacturing in the US and Europe: How digitalization will impact metropolitan areas depending on sectoral dominances and regional skill distribution

Yasmin M. Hilpert

ABSTRACT

Since digitalization and Industry 4.0 have been recognized as a key issue for future economic development, prosperity and wealth distribution, several studies have emerged on the potential threats of new technology on workforce development. The consensus is that jobs may fall away, while some new jobs will be created, with a different skills profile and a new set of qualifications that are required. This paper examines the effects of three main indicators: the impact of skills, industrial sector dominance and product complexity on workforce reduction. Based on metropolitan data from the US (Census) and Europe (Eurostat), the author develops a metropolitan typology based on industrial sectors in each metro and analyses the systematic relationship between regional variations of automation, local skills and economic sector variations, finding that automation exposure in Europe is significantly lower than in the US and that medium-skilled manufacturing jobs in the US are increasingly threatened and low-skill service jobs remain relatively safe from automation – leading to a decreasing middle class. This also shows how metropolitan areas are at risk of developing polarized effects: some facing economic upturn and continuous prosperity, and a majority of others either stagnant or with extreme downturn and high unemployment rates.

1. Introduction

In the context of the Industry 4.0 and digitalization discourse, there is a concentration on strong, leading competitive countries, changing or new technologies in industries, and the following expected impact on work, products and everyday life. However, little focus is given to the effects on regions or metropolises characterized by their particular structures and how these will be affected by the changes to come. Manufacturing industries and services are facing a fundamental change based on Industry 4.0 and general digitalization. While not all of its effects are foreseeable today, and many new effects will emerge during application, there are already certain impacts on jobs, skills and economies. This far-reaching process of innovation and application of new technological opportunities meets divergent situations in regions and, in particular, in metropolises. Thus metropolitan areas, as centres for economic development and as clusters of such industries, are becoming increasingly important. More than 74 percent of the population in the EU live in cities (World Bank 2019; compare to 84 percent in the US, Statista 2020c); cities contribute to more than 67 percent of the national GDP in Europe (Statista 2020a) and 90 percent in the US (Statista 2019). While this characterizes the differences between metropolitan areas and less urbanized areas, there are also important differences among metropolises that influence the general effects of Industry 4.0 and digitalization may have a noticeable impact. But even metropolises and regions differ amongst each other: variations in their industrial structures and the share of services, the availability of certain skills in the workforce and the complexity of products or services cause systemic variations; and such variations manifest themselves when comparing metros and regions between the US and Europe as more pronounced concentrations of manufacturing or services in the US, and more hybrid, mixed metros in Europe.

These processes of innovation and modernization are also effective in linking industrial strength, service capabilities and exchange of ideas, people and goods. Consequently, existing conditions influence adoption of such changes. In Europe and the US, metropolitan areas are gravitational centres of strong physical infrastructures like interstates, trains, airways, etc. that allow quick transport of goods and services and national and global connectivity. Leading institutions in education and research are predominantly located here and help to attract and concentrate young talent (Hilpert 2016; Trippl 2014). The share of university graduates and occupational complexity has grown continuously (Statista 2020b; Singh and Briem 2014). While they pass through their professional careers they further contribute to these situations. Companies settle in metropolitan areas for both – they need reliable infrastructures and connectivity as well as access to well-educated human capital.

Industry 4.0 and digitalization are processes that are already affecting all industries and will continue to do so. From industrial manufacturing to the service industry, all parts of our economy will experience this change in one way or another (McKinsey Global Institute 2017). Consequently, while new digital technologies change the economy, they also change metropolitan areas in most Western industrialized societies.[1] However, metropolitan areas vary along their sector dominances and the effects of digital technology may be experienced differently in a predominantly service-based metropolitan economy than in a high-tech manufacturing city. While metropolitan areas have become more important in recent decades, their development strongly depends on the dominant industries. This is particularly the case when comparing them with regard to the expected effects of digitalization and industry 4.0.[2] How metropolitan areas have developed historically and the developed industry cultures determine how these cities will evolve. New York City, for example, has always been the center of commerce and trading and continues to be strongly service-based – but looking at the greater conurbation of NYC, there is also a significant number of manufacturing firms that are drivers of industrial innovation – in absolute numbers, they concentrate more capabilities than most metros characterized by manufacturing. Similarly, Boston as well as the San Francisco Bay Area with their strong universities have developed an outstanding strength in biosciences and engineering, and Pittsburgh with its formerly strong heavy manufacturing base in the steel industry, continues to have very strong, but clearly more advanced and sophisticated manufacturing roots (Singh and Briem 2014). Similarly, in Europe, there are areas of strong manufacturing like Baden-Württemberg, Bavaria, Lombardia, Rhône-Alpes or Catalonia and centres of services such as London, Paris, Milan or Frankfurt. Metropolises clearly will face different effects of Industry 4.0 and Digitalization. This raises questions about how unique or distinct circumstances within a metropolis might affect the automation risk that those locales are exposed to and whether divergent factors in metropolises in Europe and the US finally direct towards similar tendencies but show variations in empirical data. This paper takes initial research on the automation risks in US metropolitan areas to determine whether there are sector-specific dynamics that systematically impact the risk and will also draw similar conclusions from this research for the EU, factoring in structural differences across industrial sectors between the US and the EU.

2. Automation, worker displacement and the skills misfit

2.1 Effects of digitalization and regional variations

The effects of digitalization and Industry 4.0 have been discussed broadly – both in policy and academically, but the results, projections and underlying connotations vary greatly: evaluating the potential benefits for businesses, but especially the impacts on certain professions, based on the skills and training required for them and their risk for automation and displacement, as well as the possible effects on wages and devaluation of skills, McKinsey Global Institute (2017) suggests that workers with a high school degree or less have a 55 percent chance of their jobs being automated, while those with post-secondary education are at a 44 percent risk and those with a college education are only at a 22 percent risk of losing out to automation. Maxim and Muro (2019) project a total of 25 percent of the American workforce facing high exposure risks to automation. Similarly, Barker and Berube (2018) provide an insight into how the ongoing disruptions that challenge local talent pools and training.

However, variations within countries along regions and metropolitan areas are discussed only very rarely. Most of the more prominent studies focus on nation states as unit of analysis (e.g. McKinsey Global Institute 2019; Hirsch-Kreinsen 2015, 2016).[3] As variations are strong across countries between regions and even between different metropolitan areas, the prognoses for job losses and job gains are difficult to assume as universal. Because of the differences in industrial structures, skills and research capabilities the different metropolises will not face the same situations. Similarly, a European-American comparison demands a consideration of the variations between the countries. And while cross-country comparisons can give some indication as to the existing industrial policy solutions applied in different countries to manage economic structural change in macro-level scenarios[4], on a micro-level one city may experience job growth, income growth and better employment opportunities, while another may experience job losses, declining wages and rising unemployment.

This does not, however, account for the potential systemic variations between industrial manufacturing and service industries and these potential scenarios, or variations based on existing skill levels, educational infrastructure, etc. For example, it is plausible that a specialization on high-technology production and services is the most likely scenario in Germany and several other Western and especially Northern European countries. Their strong manufacturing sector, expertise in engineering and application of new technologies and highly skilled blue collar labour force as well as the professional expertise that comes with long apprenticeships and continuous education provide a competitive advantage and an intense human capital for companies. The industrial competences formed out of the labour force of skilled and educated workers allow for specific applications and upgrades based on Industry 4.0 and digitalization. This meets a situation of regional societies strongly characterized by the highly skilled labour and the capability to cope with change in metropolitan situations. In the US, this scenario is significantly less likely – a decline in manufacturing and offshoring of medium-skill manufacturing to developing and emerging markets has caused a significant job gap and has created a vacuum that has largely been filled by a strong service sector, which employs a variety of skill levels, but a noteworthy amount of low and unskilled workers in low-wage jobs (Muro et al. 2017). Therefore, the specific effects of digitalization, even with these potential scenarios, on industries, skill levels and hence the regional variations across different countries still remain largely uncovered.

2.2 Three indicators for regionalized automation variations

The discourse on Digitalization and Industry 4.0 widely agrees that variations in effects will occur along three indicators: (1) education and training, (2) sectoral variations and (3) product or service complexity. As these three indicators vary between metropolitan areas and regions, so do the effects of digitization and automation will result in regional variations of risk and disruption.

Nonetheless, there is a controversial discussion about the effects on different parts of the workforce based on education and training. New technologies could mean that low-skill labour could be automated and replaced with modern technology if wage levels and regulation of working hours are relatively high and automation would be economically reasonable and would suit to longer times of manufacturing activity. On the other hand, digital assistance systems could also be used to help in medium-skill jobs and facilitate the task to accommodate more low-skill workers, which would allow companies to pay less based on the average skill level of their workforce (Fraunhofer-Gesellschaft 2016; Buhr 2017). Generally, the effects on different parts of the workforce according to their skill level will depend on their wages and regulations – where wages are low and regulations are weak, automation will not be economical. With the existing variations in wages between service and manufacturing industries, this puts low-skill manufacturing workers at higher risk for automation than low-skill service workers. Metropolises which are also characterized by many low-skill manufacturing jobs may face particular risks.

Medium-skill manufacturing that could benefit from digital assistance systems could simultaneously mean employment growth for low-skill workers (Hilpert 2017). The previously more training-intensive work in these industries could be supported by technology and carried out by low-skill workers. However, US metropolises may be less affected by this tendency since a large part of medium-skill (and medium-tech) manufacturing is no longer done in Western industrial societies and has been offshored to emerging markets (Vallizadeh, Muysken, and Ziesemer 2015; Holzer and Lerman 2007; Stewart 2017). It is important to note what the scale of skill is when defining high, medium and low-skill occupations: When comparing across one aggregate (e.g. the US in total), the medium skill range may be lower skill than another aggregate (e.g. EU), because when comparing across industries, these skills may be higher. For example, manufacturing work may be ranging in the medium category in the US, but in the EU, training requirements are higher (usually several years of training on the job and apprenticeships) and would therefore make a similar manufacturing industry medium-high skill in Europe.

Similarly, the effects of digitalization in metropolises are expected to occur differently because the required skill levels vary across different sectors. Metropolises reflect the fact that industries split along a line of technology development (such as mechanical engineering, information and communications technology, etc.) and technology users (such as automotive, chemical, biotech, etc.). Additionally, some industries are already heavily automated (like the automotive industry for example) and upscaling technology towards more automation and digitalization is more feasible than in other industries (Hilpert 2017). Companies at a larger scale have more financial means to invest in the new technologies to switch to smart manufacturing than small and medium enterprises. In Europe, many regions and metropolises are characterized by a significant share of SMEs which again means a lower risk of job losses due to these new technological opportunities.

Regional industrial structures are a key factor in determining how labour markets shift as a result of new technologies. Products that require little to no customization and industries with lower product or service complexity are at a higher risk of automation. Conversely, jobs and products that require human complexity and personalization face less risk of automation.

The variations along with these three basic indicators – skill level, sector and product or service complexity – also manifest geographically. While in the US this is already a politically relevant issue, in the EU, even though urban-rural divides certainly exist, the variations are not as extreme and most rural areas are somewhat nearer to larger metropolitan areas with significant economic impact on the greater region. More recently, the potential effects of digitalization and Industry 4.0 have also been analysed in relation to these strong differences between the modern, highly educated metropolitan areas and the less developed rural areas with more difficult access to higher education and training. Even more than before, metropolises will become magnets and agglomerations of skilled talent. In rural areas, digitalization and potential workforce impacts are not the only forces in place that could severely disrupt the labour force – where environmental and climate protection policies are added on top of the potential stress due to digitalization and modernization of industries, the potential effects on automation and economic downturn could be detrimental. Prognoses for national average workforce changes indicate reductions of up to 35 percent (US) and a reduction of working hours by 20 percent (McKinsey 2017). While these numbers are serious, the national data implies that variations will occur regionally and that some areas may experience even higher impacts.

As metropolitan areas vary widely depending on economic sector dominances, they vary also in their risk aversion to the workforce effects of digitalization and Industry 4.0.

2.3 The metropolitan medium skill vacuum

Both in manufacturing and the service industry, jobs requiring a medium skill level from their workers have been in decline. Outsourcing of white-collar service jobs such as copy-writing, computer programming, and some legal tasks and the rise of freelance platforms such as Upwork, Fiver, etc. have allowed for companies to cut costs by employing non-domestically, and many of the medium-skill (and often medium-tech) manufacturing has moved to emerging and developing economies in Asia (and some in Latin America) (Bishop 2012, 4). Simultaneously, targeted innovation policy and investments into research and development in many developed Western economies – take the US as an example – have increased and the required skill level has followed suit, creating the demand for an extremely skilled, specialized workforce. The paradox of the oversupply of workers and simultaneous skills shortage experienced by companies, therefore, suggests a trend that unemployment is likely to remain relatively high in developed nations even as it declines in the developing world (Bishop 2012; Sandulli and Gimenez Fernandez 2019).

Consequentially, these metropolitan societies have to face a vacuum for medium-skilled workers in their labour markets: Industry 4.0 implies an at-risk labour market in metropolitan areas with many medium-skilled workers that simultaneously does not supply the adequate number of jobs fitting these available skills. While it speaks for good policies and societal progress that the aggregate education level has risen in the past years and decades, today the lack of jobs and matching industries indicates a lack of policy focused on industry retention and strategic industrial policy planning.

The challenge of oversupply of a medium-skilled workforce is twofold: medium-skilled workers are in the dilemma of being not skilled enough to fill the gap in the high-skilled sectors and yet are too skilled for the low-skilled sectors, which pay less and offer fewer opportunities for career growth. Especially in metropolises of countries where education and training remain private, high personal debts that remain difficult to pay off without a commensurately lucrative and skilled job and present a severe challenge to the achievement of equity and social coherence.[5]

Generally, the consensus is that low-skill jobs are at high risk of automation. However, with a simple application of a cost-and-benefits calculation, it becomes clear that there are certain practical and economic conditions to this consensus that lay open the clear variations across manufacturing and service industries. There is a variation between low-skill service and manufacturing to be expected: While low-skill manufacturing is easier to automate, low-skill services are significantly less likely to be replaced and become obsolete, as customer-service related jobs, hospitality jobs, etc. are inherently cheap, rendering their automation uneconomical. This should imply that service metros with a large low-skill labour force should face lower automation risks than man-ufacturing metros with a similar skill distribution because the initial investment into Artificial Intelligence (AI) and digital technology is outweighed by the labour costs and the organization of work. Ironically, the low wages in the service industry may save them from automation and further disruption, but not from precarity.

In manufacturing, the transition to smart technologies, AI, advanced robotics and assistance systems is expensive and requires a significant upfront investment. Such investment will only be made if it is economically sensible. Small and medium enterprises, which make up ca. 99% of European companies, will not be able to make this investment upfront without considerable subsidies and government grants. At the same time, a noteworthy amount of these European SMEs themselves is in the industrial design sectors, the very drivers of automation technologies, as Europe is a strong player in producing the necessary technology to automate other industries.

Metropolises that may avoid the employment problems by several low-skilled jobs must still face the low-income characteristics of their own communities. The majority of low-skilled labour in European societies is in service industry, one of the lowest wage sectors. Automation would prove costly and uneconomical. In the US, the majority of low-skilled labour is also in the services industry, but this challenge is compounded further by the fact that the country is home to a large share of low-skilled manufacturing employees because – contrary to European societies – low wages, weak industrial relations and diminished collective bargaining power as well as over-optimized production line processes have retained some of the low-value-added manufacturing within the US. The continuous upgrading of industries and skills helps some strong economies and their metropolises avoid problems facing the US and to cope with the effects of new technologies. Given the structural differences between metropolitan areas, the variations between manufacturing and services in light of skills and workforce developments and automation threats should therefore also be visible at the metropolitan level.

3. A metropolitan typology of systematic sectoral variations on national economic development and the new digital economy – methodological remarks

The metropolitan effects of digitalization are diverse and reach beyond just a replacement of routine work and low skills. New technologies impact almost all areas of manufacturing and services. Even highly advanced economic activities employing people with high skills are affected, such as AI supportive technologies, automated systems, etc. Thus differences in the economic structure of metropolises can be identified by a typology, which helps to identify the effects of this process of innovation. This analysis demands more precise differentiation because national results will play out differently – and systematically – across regions and between metropolises. As literature shows, such differences are to be identified systematically according to effects in manufacturing industries compared to services. There are two indicators that help to build a typology: Employment by industrial sector and GDP. This differentiation is operationalized with the statistical median of both variables. It would seem random to split metropolitan areas along a simple 50 percent line – metropolitan areas that have both a very strong manufacturing and a service base may experience interactive effects that make a metropolitan area develop differently. They may vary not just about economic factors such as GPD and productivity rates, but also in their individual history of labour migration and areas of employment, as well as social determinants such as employment (especially in light of possible automation), incomes and income inequality, racial diversity and equity, etc.

Additionally to Service and Goods-Producing Metros, this analysis therefore also captures Balanced Metropolises that lie within 5 percent above and below the median (see Table 1). This is to capture the unique situation of a metropolitan area that lies between sector dominances and their specific exposure to automation and general digitalization effects.

Sectoral variances and dominances in metropolitan areas are assumed to have a strong impact on crucial variables of socio-economic development. However, depending on the size of metropolitan areas, a proportionally small manufacturing sector in a large metropolitan area could still be a very large employment generator and potentially larger in absolute terms than a smaller city that is mostly defined by manufacturing. Additional to the three types developed above, this paper considers Hybrid metropolitan areas that are defined along with absolute and proportional employment by sector (see Table 2). Among this type, Hybrid A Metros are hybrid metropolises whose employment share in the service industry are above the median and are thereby 'leaning' towards the service industry, while Hybrid B Metros have a below-median share in the service industry and are thereby 'leaning' toward goods-producing metropolises. This also allows for testing of positive interactive effects between sectors, such as an increase in manufacturing-related services in hybrid metros that could lead to employment growth compared to other metro types (such as goods producing or service metros, see Table 3).

Quantitative data for this analysis is available for NUTS2 regional level data from Eurostat in the EU and Metropolitan Statistical Areas (MSAs) in Census data from the US, which allows for cross-country or cross-continental comparisons, highlighting the variations between countries, metros and even metros of the same type, but in different countries. Since Eurostat and Census data is available for more than 600 metropolitan areas, the sample for this paper was reduced to the 100 largest metropolitan areas by population in the US and the EU respectively.

4. Examining automation risks in light of sectoral variations – analysis

While there is consensus that automation will vary across sectors and between skill levels, there is limited data to evaluate how these variations will break down regionally. Research either examines the relationship between skill, sectors and automation, or the relationship between automation and regional variations thereof. A more comprehensive attempt at the latter was done by Muro, Maxim, and Whiton (2019), who analysed the 100 largest US metros and showed that workers' educational attainment will prove most influential of automation risk. Muro et al. developed a risk categorization in three groups: automation rates between 39.8 –45.0 percent, 45.0–46.3 percent and 46.3–49.0 percent. Their data suggest that smaller, more rural communities as well as smaller metropolitan areas will be at greater risk of worker displacement.

Table 1. Typology of metropolitan areas into service, goods-producing und balanced metros.

	GDP share of service industries <45%	GDP share of service industries 45–55%	GDP share of service industries >55%
Employment share in the service industry >55%		Balanced	Service Metros
Employment share in the service industry 45–55%	Balanced	Balanced	Balanced
Employment share in the service industry <45%	Goods-Producing Metros	Balanced	

Table 2. Typology of metropolitan areas into goods-producing, service, hybrid and hybrid A & B metros.

	Employment share in the service industry (in percent) below the median	Employment share in the service industry (in percent) above the median
Absolute share of employment in the service industry below the median	Hybrid B (leaning Goods Producing) Hybrid	Service
Absolute share of employment in the service industry below the median	Goods Producing	Hybrid A (leaning services) Hybrid

To date, this work by Muro et al. is one of the only research endeavours that aim at breaking down automation risks on a regional and metropolitan level. However, Muro et al. do not account for the specific path dependencies and industry cultures in these metropolitan areas, such as the infrastructure and expertise present in each locale. The lack of more granular analysis leaves open the question of whether there are systematic variations along specific sectoral variations, skill levels, etc. and whether the proposed medium skill vacuum can be demonstrated with this data as well.

As of the time of this paper, no comparable research examines the EU workforce automation risks in the same regionalized way as Muro et al. But Muro et al. paper does allow some conclusions on the expected metropolitan variations in automation and workforce impacts in the EU if similar trends apply based on systematic sector and skill variations.

For this paper, a cluster analysis was performed to develop five categories among the top 100 largest US metropolitan areas[6] based on the percentages of employment in low-, medium- and high-skill occupations. These categories can loosely be labelled as low-skill metros, medium-low skill metros, balanced skill metros (neither skill level was particularly outstanding in frequency), medium-high skill metros and high skill metros (see Table 4). A logistic regression run on the dependent variable of the automation percentage from Muro, Maxim, and Whiton (2019) shows that, indeed, a higher skill level corresponds to a lower automation risk level (see Table 5): Metros with a particularly high-skilled workforce (employment in high-skill occupations) have a ca. 4 percent lower risk of automation than those with a low-skilled labour force.

This data supports the general consensus that skill level is an important predictor for automation risk, even at the microlevel when taking a closer look at metropolitan areas. The picture becomes even clearer when the metro types are examined to the automation risk (see Table 6): Of the 20 goods-producing metros in the top 100 largest metropolitan areas, 13 are in the high automation risk category, while only 6 out of 27 of the service metros are at high risk of automation. This supports the argument that where wages are low, such as in many service jobs, automation may not be economical, while the formerly

Table 3. Frequency distribution of generated metropolitan types.

Metropolitan type	Islands of innovation	Top 10%	Total
Goods-Producing	1	8	25
Service-Providing		7	28
Balanced	2		10
Hybrid A (leaning Service-Providing)	8		17
Hybrid B (leaning Goods-Producing)	6		20
Total	17	15	100

Table 4. Results of a KNN5 cluster analysis for employment in low, medium and high skill occupations in the Top 100 largest US metropolitan areas.

LABEL	KNN5 CLUSTER CODE	AVERAGE % LOW SKILL OCCUPATION	AVERAGE % MED. SKILL OCCUPATION	AVERAGE % HIGH SKILL OCCUPATION
Low	4	45.8	36.6	17.7
MEDIUM–LOW SKILL	1	32.5	47.9	19.7
BALANCED SKILL	2	30.6	39.2	30.2
MEDIUM–HIGH SKILL	5	23.8	45.7	30.5
HIGH SKILL	3	25.4	33.9	40.7

Data sources: US census data 2016, own calculation.

classic middle-class jobs in manufacturing, such as assembly line work at an automobile manufacturer, are at much higher risk of automation.

Between the US and EU data, it is important to note that education levels are expected to be higher, as most of the medium skill jobs in the US require a significantly shorter apprenticeship or training on the job (usually between 6 and 12 months) than in most EU countries (usually between 2 and 4 years). This should result in a lower automation risk in the EU, as the entire cohort ranges higher in education and training requirements to fulfil the skills necessary for the observed occupations.

Additionally, the US manufacturing sector (country data) as of 2017 accounts for 11.4 percent of GDP (compared to 77 percent in services), while Germany's manufacturing sector accounts for 26.8 percent (compared to 62.4 percent in services) (Statista 2020c). This implies that the effects of industry 4.0 and disruptive technology – automation risk – are exacerbated by the strong service sector and supported by the variation in wage distribution between service and manufacturing metros in the US and Europe. While the variables for household income in the US and Europe do not allow for a cross-sample comparison, it is possible to compare within sample and between metro types: In the US, the household income in service metros is almost 5 percent lower than in goods-producing metros, while in the EU, disposable household income in service metros is almost 16 percent higher than in good producing metros. If we know that automation of jobs will only happen where (a) it is economical and wages are high enough to make automation profitable and where (b) skills are in the medium range, then this further illustrates that in the US, manufacturing jobs are at high risk of automation.

Meanwhile, hybrid metros, especially service-providing hybrids, as well as Islands of Innovation have the lowest risk of automation. It is worth noting that at 39.8 percent

Table 5. Results of a logistic regression for the impact of occupational skill level on automation risk on the metro level (Top 100 largest US metros).

VARIABLE	DEPENDENT VARIABLE: AUTOMATION RISK			
	ESTIMATE	ERROR	T-VALUE	PR > (|T|)
MEDIUM-LOW SKILL	−1.33 (.)	0.77	−1.72	0.088581
BALANCED SKILL	−2.50 (***)	0.67	−3.74	0.000323
MEDIUM-HIGH SKILL	−2.09 (**)	0.66	−3.15	0.002242
HIGH SKILL	−4.00 (***)	0.72	−5.54	3.03e-07

Data source: US census data 2016, own calculation.

Table 6. Contingency table of US metro type and automation risk group.

Metro type	Automation level 39.8–45%	Automation level 45–46.3%	High Automation level 46.3–49%	Total
Goods producing	4	3	13	20
Service providing	10	11	6	27
Balanced	3	3	4	10
Hybrid A (service)	12	3	2	20
Hybrid b (goods)	4	9	7	17
Hybrid (total)	16	12	9	37
Top 10 Goods	5	-	-	5
Top 10 service	4	3	-	7
Islands of innovation	11	4	2	17

Data source: US census data 2016, Muro et al. (2019), own calculation.

minimum automation (see Table 7), this cannot really be understood as a truly low risk, but rather the lowest category in the dataset. A workforce displacement at this level would have tremendous effects on welfare systems and society as a whole and poses a huge challenge to policymakers in an effort to ensure social cohesion and prosperity.

The superior performance of hybrid metros and in particular the Islands of Innovation is a continuation of their general outstanding in the data (see Tables 8 and 9). Across many innovation indicators, they are true drivers of growth, invention and global competition. Their competitive advantage is an extremely skilled workforce, by which they outperform every other metro type in many ways: In the US, they are 6 percent above the average for highly skilled employment, they contribute more than 50 percent above average to the total number of patents and their GDP contribution is exceptionally high, while their share of the population living under the poverty line is also significantly lower. In Europe, while all tendencies and comparisons between metro-politan types lean in the same direction as the US, the variations are significantly lower

Table 7. Average comparison across metro types in the EU and the US.

	Poverty			Education (Share Tertiary, excl. Associate Degree)			Household Income*	
	Mean EU	Mean US	p	Mean EU	Mean US	p	Mean EU	Mean US
All Cases	26.763	40.023	<0.001	31.387	31.063	0.796	15,065.06	95,945.77
Balanced	27.014	35.713	0.032	30.93	35.193	0.207	14,966.67	106,377.10
Goods-Producing Hybrid	18.145	37.573	<0.001	27.544	33.213	0.03	17,575,00	102,266.60
Service-Providing Hybrid	27.425	37.401	0.063	32.883	35.146	0.51	16,372.73	104,340.10
Service-Providing	34.0571	44.564	0.13	38.158	28.481	<0.001	15,422.22	86,951.50
Goods-Producing	29.207	40.406	0.002	25.341	27.805	0.202	12,993.10	91,081.98
Islands of Innovation	19.327	34.611	<0.001	36.111	37.485	0.567	18,841.00	112,901.70
Hybrid	22.053	37.494	<0.001	29.832	34.102	0.04352	17,085.19	103,219.30
Goods-Producing Top 10%**	39.183	40.641	0.68	18.411	24.706	0.033	8,855.56	89,737.24
Service-Providing Top 10%**	27.967	42.304	0.244	42.05	28.881	0.017	16433.33	90643.1

* This variable is not comparable between the two samples of the US and Europe as the US census captures the average household income, while Eurostat captures the disposable household income. It should only be compared within samples and between metro types (e.g. within the EU, between goods producing and service providing metros).
** Due to the small sample size, this group should not be compared.
Data source: US census data, Eurostat, own calculation.

Table 8. Performance of US metro types across innovation indicators (Top 100).

Metro type	Share Manufacturing Employment	Share Service Employment	Share of Employees in Low-skilled Occupations	Share of Employees in Medium-skilled Occupations	Share of Employees in High-skilled Occupations	Patents per 100K Population	GDP in Million USD	Share of Population with Less than 60% of the National Average of Disposable Income
Year	2017	2017	2017	2017	2017	2011	2016	2017
Good producing	23.46 (5.27)	76.54 (−5.27)	29.91 (1.57)	42.5 (0.94)	27.59 (−2.51)	34.78 (−4.55)	34,021.04	40.41 (0.39)
Service providing	14.25 (−3.94)	85.75 (3.94)	29.64 (1.3)	43.9 (2.34)	26.47 (−3.64)	22.35 (−16.98)	448,84.39	44.56 (4.54)
Balanced	17.75 (−0.44)	82.25 (0.44)	26.8 (−1.53)	40.28 (−1.28)	32.92 (2.81)	45.19 (5.86)	209,728.5	35.71 (−4.31)
Top 10 goods producing	27.44 (9.25)	72.56 (−9.25)	32.32 (3.99)	41.68 (0.12)	26 (−4.11)	27.6 (−11.73)	31,831.06	40.64 (0.62)
Top 10 service	10.75 (−7.44)	89.25 (7.44)	28.18 (−0.16)	43.71 (2.15)	28.12 (−1.98)	24.36 (−14.97)	439,00.31	42.31 (2.29)
Hybrid	17.74 (−0.45)	82.26 (0.45)	26.75 (−1.58)	39.58 (−1.98)	33.66 (3.56)	53.68 (14.35)	257,480.6	37.49 (−2.53)
Hybrid service	13.84 (−4.35)	86.16 (4.35)	27.16 (−1.18)	39.08 (−2.48)	33.77 (3.66)	45.2 (5.87)	338,599.4	37.4 (−2.62)
Hybrid manufacturing	21.05 (2.86)	78.95 (−2.86)	26.41 (−1.93)	40.03 (−1.53)	33.56 (3.46)	60.89 (21.56)	188,529.7	37.57 (−2.45)
Islands of innovation	17.37 (−0.82)	82.63 (0.82)	26.62 (−1.71)	37.35 (−4.21)	36.03 (5.92)	90.45 (51.12)	427,981	34.61 (−5.41)

Data source: US census data 2016, own calculation.
Numbers in brackets are percentage points above or below the average for all metropolitan areas.

Table 9. Performance of EU metro types across innovation indicators (Top 100).

Metro type	Share manufacturing employment	Share service employment	Share of employees in low tech industries	Share of employees in medium tech industries	Share of employees in high tech industries	Patents per 100K population	GDP in million EUR	Share of population with less than 60% of the national average of disposable income
Year	2017	2017	2017	2017	2017	2011	2016	2017
Good producing	19.72 (4.96)	59.64 (−11.37)	41.75 (0.72)	53.19 (1.72)	5.45 (−1.66)	6.67 (−4.64)	40956.17	23.42 (1.75)
Service providing	8.79 (−5.97)	79.25 (8.24)	37.94 (−3.09)	52.56 (1.09)	8.17 (1.06)	11.96 (0.65)	84446.15	18.3 (−3.37)
Balanced	14.52 (−0.24)	73.24 (2.23)	43.79 (2.76)	49.9 (−1.57)	6.06 (−1.05)	12.1 (0.79)	98587.6	19.37 (−2.3)
Top 10 goods producing	21.39 (6.63)	46.02 (−24.99)	39.9 (−1.13)	57.03 (5.56)	4.05 (−3.06)	0.22 (−11.09)	20462.67	26.27 (4.6)
Top 10 service	5.2 (−9.56)	84.69 (13.68)	37.45 (−3.58)	50.73 (−0.74)	9.7 (2.59)	15.62 (4.31)	108656.5	30.5 (8.83)
Hybrid	16.76 (2)	72.28 (1.27)	42.93 (1.9)	49.17 (−2.3)	7.7 (0.59)	15.15 (3.84)	165006.8	20.97 (−0.7)
Hybrid service	10.63 (−4.13)	79.29 (8.28)	41.14 (0.11)	49.5 (−1.97)	8.98 (1.87)	11.99 (0.68)	196640.4	23.95 (2.28)
Hybrid manufacturing	21.35 (6.59)	67.03 (−3.98)	44.27 (3.24)	48.93 (−2.54)	6.74 (−0.37)	17.52 (6.21)	141281.5	15 (−6.67)
Islands of innovation	13.28 (−1.48)	76.69 (5.68)	38.72 (−2.31)	51.29 (−0.18)	8.83 (1.72)	21.4 (10.09)	179375.9	15.95 (−5.72)

Numbers in brackets are percentage points above or below the average for all metropolitan areas.
Data source: US census data 2016, own calculation.

	GDP share of service industries <45%	GDP share of service industries 45-55%	GDP share of service industries >55%
Employment share in the service industry >55%		Balanced	Service Metros
Employment share in the service industry 45-55%	Balanced	Balanced	Balanced
Employment share in the service industry <45%	Goods-Producing Metros	Balanced	

Figure 1.

	Employment share in the service industry (in percent) below the median	Employment share in the service industry (in percent) above the median
Absolute share of employment in the service industry below the median	Hybrid B (leaning Goods Producing) Hybrid	Service
Absolute share of employment in the service industry below the median	Goods Producing	Hybrid A (leaning Services) Hybrid

Figure 2.

and point to a previously active industrial policy that has made variations between different metro types less noticeable.[7]

What remains true is that metropolitan areas are structurally different along systematic cleavage lines, specifically their economic sector dominances. A comparison of averages per metro type (see Table 7) shows just that. For example, poverty rates in

Metropolitan Type	Islands of Innovation	Top 10%	Total
Goods-Producing	1	8	25
Service-Providing		7	28
Balanced	2		10
Hybrid A (leaning Service-Providing)	8		17
Hybrid B (leaning Goods-Producing)	6		20
Total	17	15	100

Figure 3.

service metros are significantly higher than in goods-producing metros (both in the US and in the EU). But it also shows that despite the same methodology for the typology being applied in the EU and the US, the specific path dependencies (such as decades of industrial and educational policy) in both comparison groups have led to strong structural differences. This becomes very clear when comparing the averages within typologies but across the EU and the US: Poverty rates in all metro types in the US are significantly higher than in the EU, even in those metro types that are developing relatively well. In balanced, hybrid metros and Islands of Innovation (which have the lowest poverty rates within the US), they are still significantly higher ($p < 0.001$) than in the same metro types in Europe (balanced metros have an 8 percent higher poverty rate, and for hybrid metros and Islands of Innovation the difference between the EU and the US is 15 percent). The highest poverty rates in the US and in the EU are found in service providing (at 44 percent in the US, 34 percent in the EU) and good producing metros (40 percent in the US, 29 percent in the EU). The added stress of automation, which was found to be particularly intense in the US for the manufacturing sector, will therefore further challenge the already precarious situation of many American working families while in Europe, these effects are less likely to be as extreme because poverty rates are lower, social welfare is better equipped to cushion the effects of economic and industry change and vocational training levels are higher.

It is also worth noting that the metros in the US that rank highest in poverty rates, specifically service-providing metros, are also those with the highest average education level. Between the metro types, however, and when comparing between the US and the EU, this data point only captures tertiary education and not the level of vocational training. This data was not available in the Census or Eurostat data samples, but the effects of longer vocational training on job security, even in light of technology disruptions are well known – as is the case for the differences in length of vocational training and apprenticeships in the US compared to other Western economies.

5. Conclusion

The data analysed in this paper has provided an insight into the systematic relationship between economic sectors, skills and automation risk. It has shown that in the light of the outstanding role of metros and their contributions to national economic development, they face drastically different challenges, not just because of the automation risk and worker displacement, but because of the systematically different circumstances that these metropolises are confronted with, depending on their economic structure, sector dominances and human capital. We have also observed the outperformance of hybrid and balanced metros and in particular the Islands of Innovation in regard to many classic innovation indicators. It is worth noting that these differences between the metro types are particularly strong in the US – the comparison with the EU metros that were classified with the same methods, show the same trends, but they are not as clear as in the US.

A superficial conclusion could be that the innovation hotspots that these Islands of Innovation represent, ought to be the model and focus of policy makers because they seem to be the most resilient and socially equitable. However, this conclusion would overlook that Islands of Innovation are relatively homogenous in regard to the skill level of their workforce. A highly skilled workforce resulting in a lower automation risk and a relatively equitable labour market and lower poverty rate is therefore not a surprising finding. The much more pressing conclusion of this paper is this: The outperformance of some metro types (hybrid, balanced and Island of Innovation) paint a clear picture of the pitfalls of many of the other metropolises and their vulnerabilities, specifically the lack of training (and training abilities), a continuous lack of coherent industrial policy to provide resiliency of labour markets and a continuation of low wages, especially in the service industry, that leave workers employed, but seldom gainfully so.

Highlighting the variations between metros, it is also important to note the variations among metros of the same type between the US and EU. As a general equity variable, the share of population with less than 60 percent of the national average disposable income is an important metric and shows that across all metropolitan types, the EU ranges between 14 (top 10 percent metros strongest in manufacturing sector) and 26 percentage points (service providing metros) lower than in the US. The comparison with the EU data, where labour relations are more established, social dialogue and continuous education and training are accepted and heavily subsidized – and in many of the Western and Northern European countries even written into their respective federal constitutions – shows that these long term industrial policy structures and instruments through long-term industrial policy planning, welfare and education policy haven proven successful in keeping the automation risk significantly lower, not only when evaluating the prognosis, but also how the transition to a smart manufacturing economy is politically planned (such as with just transition programs etc.).

Notes

1. A citizen-oriented evaluation of the effects of digital infrastructure and artificial intelligence in metropolitan areas can be found in Tomer (2019), including the many ways in which data circulation affect and facilitate human decision making, privacy concerns and an entirely new facet to structural inequity biases that are enforced and intensified by technology on an everyday basis.
2. A more comprehensive industry and occupation-based analysis of the effects of digitalization is Muro et al. (2017). The report evaluates the risk of jobs, occupations and specific industries and found that while digitalization has been an ongoing trend in the US, the economy has experienced especially strong digitalization trends in the past 10 years and is continuously accelerating.
3. See also Lin, Shyu, and Ding (2017), Seet et al. (2018), Tessarini and Saltorato (2018), Ehlers (2020).
4. See, for example Buhr (2017), who lays open three potential automation scenarios with different worker displacement consequences for the German economy.

5. A simple comparison of the average personal debts for adults in the US with private and Germany with public higher education shows the drastic impact: with more than USD 90,000 per adult in the US, Germans only carry about USD 30,000 of debt (Experian 2020).
6. As the concrete research methods of the paper from Muro, Maxim, and Whiton (2019) were not available, there is a discrepancy of 5 metropolises between their generation of the top 100 list and the one used in this paper. For this reason, the author will use the thereby generated top 95 list for the analysis going further.
7. 0.6 percent above average for high-tech employees and 3.8 percent above average for patents.

Disclosure statement

No potential conflict of interest was reported by the author(s).

References

Barker, R., and Berube, A. 2018. Delivering Shared Prosperity for Workers in a Rapidly Changing Economy. Online accessible at: https://brook.gs/2QgMEUj (last accessed on March 12, 2021)

Bishop, M. (2012). The Future of Jobs. The Great Mismatch. Special Report from *The Economist*, 48 p.

Buhr, D. (2017). Soziale Innovationspolitik für die Industrie 4.0. Friedrich Ebert Foundation WISO Special Issue. Online accessible at: https://bit.ly/3ssnwHi (last accessed on March 12, 2021)

Ehlers, U. 2020. *Future Skills. Lernen der Zukunft, Hochschule der Zukunft*. Karlsruhe: Springer VS.

Fraunhofer-Gesellschaft. 2016. Better Quality Control with Digital Assistance Systems. Online accessible at: https://bit.ly/3dvMAZA (last accessed on March 12, 2021)

Hilpert, U. 2016. "Metropolitan Locations in International High-Tech Networks: Collaboration and Exchange of Creative Labour as a Basis for Advanced Socio-Economic Development." In *Routledge Handbook of Politics and Technology*, edited by U. Hilpert, 281–298. Abington, Oxon/New York, NY: Routledge.

Hilpert, Y. (2017): *Dramatic Changes. The Role of Unions for the Future of Modern Societies in the Light of Structural Diversities. The Challenge of Industry 4.0 and the Demand for New Answers. White Paper for IndustriAll Global Union.* Geneva, Switzerland: IndustriALL Global Union.

Hirsch-Kreinsen, H. 2015. *Digitalisierung Industrieller Arbeit: Die Vision 4.0 und Ihre Sozialen Herausforderungen*. Baden-Baden: Nomos/Edition Sigma.

Hirsch-Kreinsen, H. 2016. "Digitalization of Industrial Work: Development Paths and Prospects." *Journal for Labour Market Research* 49 (1): 1–14. doi:10.1007/s12651-016-0200-6

Holzer, H., and R. Lerman. 2007. *America's Forgotten Middle-Skill Jobs. Education and Training Requirements in the Next Decade and Beyond*. Washington, DC: The Urban Institute. Online accessible at: https://urbn.is/3edWRJ1 (last accessed on March 12, 2021)

Lin, K., J. Shyu, and K. Ding. 2017. "A Cross-Straight Comparison of Innovation Policy Under Industry 4.0 and Sustainability Development Transition." *Sustainability* 9 (5): 786. doi:10. 3390/su9050786

Maxim, R. and Muro, M. 2019. Automation and AI Will Disrupt the American Labor Force. Here's How We Can Protect Workers. Online accessible at: https://brook.gs/3n0BBKF (last accessed March 12, 2021)

McKinsey Global Institute. 2017. *Jobs Lost, Jobs Gained – Workforce Transitions in a Time of Automation*. San Francisco/Amsterdam/Shanghai: MGI. 28 p.

Muro, M., S. Liu, J. Whiton, and S. Kulkarni. 2017. *Digitalization and the American Workforce*. Washington, DC: The Brookings Institution. 60 p.

Muro, M., R. Maxim, and J. Whiton. 2019. *Automation and Artificial Intelligence: How Machines are Affecting People and Places*. Washington, DC: The Brookings Institution. 108 p.

Sandulli, F., and E. Gimenez Fernandez. 2019. "Underemployment of Middle-Skilled Workers and Innovation Outcomes. A Cross-Country Analysis." In *Diversities of Innovation*, edited by U. Hilpert, 137–153. Abington, Oxon/New York, NY: Routledge.

Seet, P., J. Jones, J. Spoehr, and A. Hordacre. 2018. *The Fourth Industrial Revolution: the Implications of Technological Disruption for Australian VET*. Adelaide, SA: NCVER. Online accessible at: https://bit.ly/3tvk5k9 (last accessed March 12, 2021)

Singh, V., and C. Briem. 2014. "Metropolitan Area Migration Patterns of the Scientific and Engineering Workforce in the United States." In *Networking Regionalised Innovative Labour Markets*, edited by U. Hilpert, and H. Lawton Smith, 78–95. Abington, Oxon/New York: Routledge.

Stewart, F. 2017. *The STEM Dilemma. Skills That Matter to Regions*. Kalamazoo, MI: W.E. Upjohn Institute for Employment Research. 207 p.

Statista. 2019. The Top 25 Metro Areas Make Up Half of U.S. GDP. Online accessible at: https://bit.ly/3amvJ9y (last accessed March 12, 2021)

Statista. 2020a. European Cities by GDP. Online accessible at: Online accessible at: https://bit.ly/2P1azqf (last accessed March 12, 2021)

Statista. 2020b. Percentage of U.S. Population with Completed Four or More Years of College Education, 1940–2019, by Gender. Online accessible at: https://bit.ly/3uZIWNx (last accessed March 12, 2021)

Statista. 2020c. Germany: Distribution of Gross Domestic Product (GDP) across Economic Sectors from 2009 to 2019. Online accessible at: https://bit.ly/32oowSb (last accessed March 12, 2021)

Tessarini, G., and P. Saltorato. 2018. "Impactos da Indústria 4.0 na Organização do Trabalho: Uma Revisão Sistemática da Literature." *Revista Produção Online* 18: 743–769. doi:10.14488/1676-1901.v18i2.2967

Tomer, A. 2019. Artificial Intelligence in America's digital city. Online accessible at: https://brook.gs/3×1HKLz (last accessed March 12, 2021)

Trippl, M. 2014. "Star Scientists, Islands of Innovation and Internationally Networked Labour Markets." In *Networking Regionalised Innovative Labour Markets*, edited by U. Hilpert, and H. Lawton Smith, 58–77. Abington, Oxon/New York: Routledge.

Vallizadeh, E., Muysken, J. and Ziesemer, T. 2015. Offshoring of Medium-Skill jobs, Polarization, and Productivity Effect. Implications for Wages and Low-Skill Employment. Institute for Employment Research. IAB Discussion Paper 7/2015. Online accessible at: https://bit.ly/3uYpsZG (last accessed March 12, 2021)

World Bank. (2019). Urban Population – United States. Online accessible at: https://bit.ly/3tBl5DK (last accessed March 12, 2021)

The Korean approach to Industry 4.0: the 4th Industrial Revolution from regional perspectives

Sunyang Chung and Jiyoon Chung

ABSTRACT
Today the concept of Industry 4.0 has been widely adopted by many countries. It is rather a narrow concept, compared to that of 4th Industrial Revolution, which presumes a wide impact on the national economy and society. The Korean approach to the Industry 4.0 is the 4th Industrial Revolution. The Korean government established the 'Presidential Committee on the Fourth Industrial Revolution' (PCFIR) and has initiated major agendas related to the Industry 4.0. One of the 'Committee's' major agendas is the diffusion of smart factories, which play an important role in the Industry 4.0. However, the Korean approach to the 4th Industrial Revolution has been difficult to be successfully implemented because regional governments have not participated in the 'Presidential Committee'. Smart factories have diffused unevenly since Korea's R&D potential, which needed for their adoption, have been concentrated in Seoul, Gyeonggi and a few industrialized regions. Korea needs to take these points into consideration in its implementation of the 4th Industrial Revolution in the future.

1. Introduction

As information and communication technologies (ICTs) advance and diffuse widely in our industries and society, the digitalization of industrial production has been bolstered. Industrial production has been automatized, and manufacturing companies and industries gained more production efficiencies and competitive advantages. In some countries, digitalization is not restricted to industrial production but embraces the national economy and society.

In this trend, researchers have engaged in a series of extensive discussions on digitalization. Such discussions have focused on the Industry 4.0 and the 4th Industrial Revolution. While the former rather concentrates on the digitalization of industrial production, i.e. the manufacturing sector, the latter is a rather broader concept that deals with the impact of digitalization on society and the economy. Today, two terms have been used interchangeably depending on the structural conditions of nation and society.

However, both concepts are not new. At the beginning of the last century, a Russian economist Kondratieff (1935) argued that Western economies had experienced a series of long waves about 50 years after the 1st Industrial Revolution. In the 1970s, neo-Schumpeterian economists emphasized that the long waves have been based on technological revolution, we are experiencing the 5th long wave, and nations as well as firms should prepare for coming long waves effectively (Freeman and Perez 1988; Perez 2002). As these three terms have similarities and differences, we might need a more detailed discussion on them.

Many countries have had strong interests in the digitalization of industrial production and their national economies. This was especially the case for the countries which have manufacturing-oriented industrial structures such as Germany, Italy, Japan and Korea. For example, Germany, which has traditionally secured remarkable competitive advantages in manufacturing sectors, has introduced the programme called Industry (Industrie) 4.0 as a main part of the 'High Technology Strategy (Hightech-Strategie) Program' since 2006 (BMBF 2016a). Many countries have benchmarked the German Industry 4.0 and Korea is one of them.

The digitalization of industrial production has been based on the widespread development and diffusion of ICTs in manufacturing sectors. This is the Industry 4.0, which is a narrow concept. However, the 4th Industrial Revolution emphasizes that digitalization has a widespread influence on our society. Schwab (2016) argued that we are at the entrance of the 4th Industrial Revolution, which would have a widespread influence on every aspect of our society, and we should seriously prepare for the Revolution.

Historically Korea has developed through the regional concentration of industrial and technological competences, which was an important element of national development strategy, especially in the 1970s and 1980s. Based on this strategy, Korea could attain high-level capabilities in major industrial and S&T areas (Kim 1997; Chung 2003, 2016, 2019). It has accumulated world-class ICT capabilities, which would be helpful for Korea's digitalization.[1] Korean big companies such as Samsung Electronics and LG Electronics have actively participated in the development and commercialization of ICTs and digital technologies. Most of big Korean ICT companies are in Gyeonggi Province, the outskirts of Seoul, and have made a significant contribution to the concentration of ICT and related industries in this region.

Regarding the digitalization and Industry 4.0, Korea has adopted a total mobilization system. The current government recognized the importance of digitalization, it has asserted that Korea should prepare for the 4th Industrial Revolution as effectively as possible, and the preparation has become a major national agenda. Therefore, the Korean government has established the 'Presidential Committee on the 4th Industrial Revolution (PCFIR)' in 2017 and all ministries have ought to prepare for the 4th Industrial Revolution and initiated a series of policy actions for it.

However, it seems that Korea has difficulties in preparing for and responding to the 4th Industrial Revolution because its' technological and R&D capabilities are not evenly distributed over Korean regions. Korean R&D and innovation capabilities have been concentrated heavily in Gyeonggi, Seoul and in the southeastern industrialized regions such as Gyeongbuk and Gyeongnam. This concentration issue might hinder Korea's preparation for the 4th Industrial Revolution. In fact, one of the weaknesses of the Korean national innovation system is the uneven distribution of R&D, technological and innovation capabilities among Korean regions (Chung 2002; Hwang, Cheong, and Chung 2019; MSIT 2019).

Therefore, it would be very interesting to know how effectively Korea can prepare for and respond to the 4th Industrial Revolution from the perspective of regions. According to the 'I-Korea 4.0', which is the 'Masterplan for Korea's Preparing for the 4th Industrial Revolution', there are 12 concrete technology-oriented areas that Korea aims to address (PCFIR 2017). However, due to the regional disparity of technological and R&D potentials among regions, it seems to be difficult for Korea to effectively deal with these areas.

The purpose of this paper is to investigate how effectively Korea could prepare for and respond to the 4th Industrial Revolution from regional perspectives. The paper is organized as follows. In Section 2 after the Introduction in Section 1, we will investigate the similarities and differences among Long Wave Theory, Industry 4.0 and 4th Industrial Revolution, that are confusingly used in all over the world. In Section 3, we will analyse the major contents of the Korean approach to the Industry 4.0. from regional perspectives. In particular, we will investigate the role of the 'Presidential Committee on the 4th Industrial Revolution (PCFIR)' Section 4 will be a case analysis on the Korean approach to the 4th Industrial Revolution, Here the Korean Program for Diffusing and Advancing Smart Factories will be analysed from regional perspectives. The diffusion of smart factories is one of the major promotion areas of the PCFIR. Finally, based on the theoretical and empirical analyses above, we will draw some policy conclusions which might be of interest for international readers.

2. Some literature reviews

2.1. 4th Industrial Revolution and Industry 4.0

New technologies permeate every aspect of human life. The impact of new technologies is widespread, and it is much stronger and broader than the 1st Industrial Revolution in the late phase of the eighteenth century. Therefore, there have been many studies on the influence of technologies on economic and social development. At the beginning of the last century, Schumpeter (1911) emphasized that technological innovation is a driving force for business cycle, economic and social development. According to him, a capitalistic economy experiences a long business cycle of about every 50 years due to technological innovation. The long business cycle is called Long Wave. As mentioned, Kondratieff (1935) emphasized that Western economies had been experiencing a series of long waves since the 1st Industrial Revolution. Schumpeter (1939) names the long waves as Kondratieff waves in honour of him. His theories and arguments have been widely accepted by many experts since the 1970s, who are called neo-Schumpeterian economists. According to them, representatively Freeman and Perez (1988), Western countries were in the middle of the 5th Long Wave as of the beginning of the 1990s.

One representative study on industrial revolution is the work by Rifkin (2011) who named the 3rd Industrial Revolution. In this book, he emphasized that the world would be dramatically changed due to the rapid diffusion of Internet technology and renewable energy. According to him, the 3rd Industrial Revolution will create different businesses and jobs and dramatically change the society, government, business, education and so on. However, the most influential study is probably the work by Schwab

(2016). He argued that the 4th Industrial Revolution has been occurring since the middle of the last century and it is characterized by the convergence of technologies that is blurring the lines between the physical, digital and biological spheres (Schwab 2016). However, it seems that the concept of 4th Industrial Revolution is too broad and ambiguous to be implemented in industrial practices and regional development.

The concept of the 4th Industrial Revolution seems to be broader than that of Industry 4.0. The former is concerned with the broader influences of ICT technologies on almost every industry and society and many discussions have concentrated on how a nation should prepare for the Revolution. However, the latter tends to deal with the influences of ICT technologies on the manufacturing sector and its discussion has focused on how the adoption or diffusion of digital technologies can improve the efficiency or productivity of manufacturing and some service sectors. For example, Carvalho et al. (2018) argue that the Industry 4.0 as a new and comprehensive industrial development concept can make a significant contribution to sustainable industrial development. Based on its high-level digitalization, virtualization and integration, the Industry 4.0 can minimize industrial waste, improve the efficient use of raw materials and natural resources, and enhance the efficiency of energy consumption and so on.

Neugebauer et al. (2016) propose the Fraunhofer Industrie 4.0 Layer Model to understand and utilize Industrie 4.0 in the context of Fraunhofer Society, which is one of the largest applied research organizations in Europe. Focusing on industrial production, Figure 1 shows the three layers which are closely interrelated with each other:

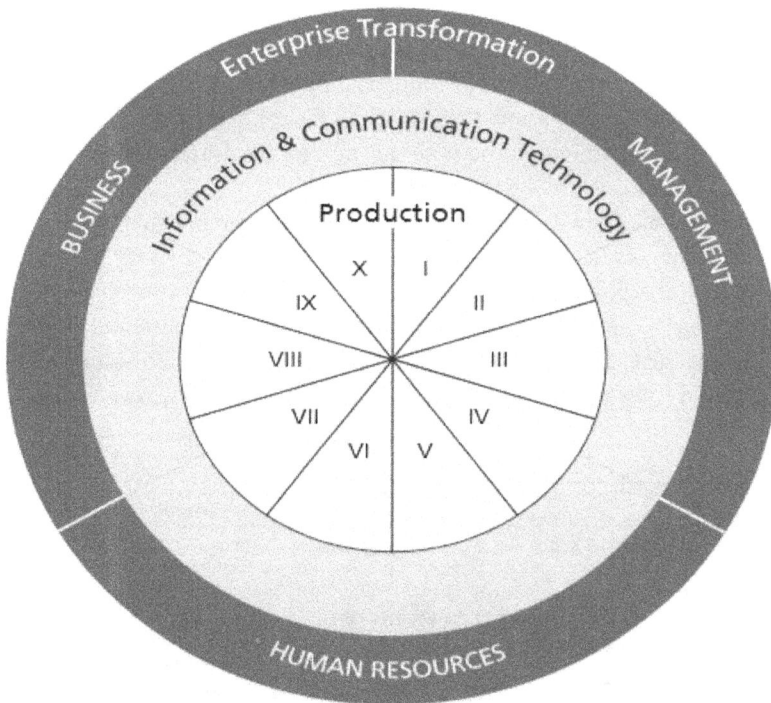

Figure 1. Fraunhofer Industrie 4.0 Layer Model. (Source: Neugebauer et al. 2016).

(1) Outer layer: human resources management, business models, business planning and cases, and transformation and change management
(2) Middle layer: ICT-enabling technologies
(3) Core layer of production: engineering, manufacturing technologies and organization, machines, smart capabilities, robotics and human-robot collaboration, production planning and control, logistics, work organization, workplace design and assistance, resource and energy efficiency

The Fraunhofer model emphasizes that industrial production is the core base of Industry 4.0. However, this proposition raises a question whether industrial production is regionally well distributed to attain the goal of Industry 4.0. Regardless, many scholars view Industry 4.0 and 4th Industrial Revolution as similar and use them interchangeably. The preference for the two terms depends on countries. Germany prefers Industry 4.0, while Asian countries such as Korea, Japan and China tend to use the term of 4th Industrial Revolution. It seems that these Asian countries are very centralized and prefer to adopt a centralized and broad approach by the centralized government. However, Neugebauer et al. (2016) argue that the Industry 4.0 is the German version of the 4th Industrial Revolution and that the term was created to emphasize the opportunities of digitalizing and integrating all instances of value-adding system, especially in industrial production. This argument suggests that both concepts have many similarities.

2.2. Comparison of the Industry 4.0-Related Concepts

We can summarize some characteristics of three related terms in Table 1. They differ from each other in their histories, characteristics and preferences by countries. In general, Asian countries prefer the 4th Industrial Revolution, while European countries prefer the Industry 4.0. Even though these three terms could be used interchangeably, but the responding mechanism would be different due to the differences in their meanings and scopes.

The concept of 4th Industrial Revolution is very comprehensive and thus is rather ambiguous and difficult to implement concretely. Therefore, it seems that this concept has been adopted at a policy level to prepare for the future. By contrast, the concept of Industry 4.0 is very concrete, given its orientation towards production, industries and manufacturing sectors. Many countries have implemented this concept and industrial companies have a long history of adopting many related technologies such as flexible

Table 1. Comparison of the Industry 4.0-related concepts.

Concepts	History	Characteristics	Preferring countries
Long wave theory	Since the beginning of the twentieth century	- Historic - Systematic - Academic	Almost all countries
Industry 4.0	Since the beginning of the twenty-first century	- Industry-oriented - Very concrete - Very practical	Germany UK France
4th Industrial Revolution	Since 2016	- Policy-oriented - Rather ambiguous - Very comprehensive	Korea China Japan

manufacturing systems (FMS) and computer integrated manufacturing (CIM) since the beginning of the 1980s (Gold 1989; Lay 1993; Chung 1996). Perhaps we can understand that the Industry 4.0, which is a relatively narrow concept, is part of the 4th Industrial Revolution.

This kind of relationship could apply to Korea because the Korean version of Industry 4.0 in a narrow sense is an important part of the Korean approach to the 4th Industrial Revolution. However, the regional disparity of Korea's industrial and technological capabilities could be a significant bottleneck against adopting the Korean approach. For example, Korean regions such as Seoul, Gyeonggi and southeastern regions of South Korea have a strong industrial and technological capabilities, while southwestern regions, which are largely agricultural, have weak capabilities.

The common implication of these three concepts is that any country and company should effectively prepare for the changes of industrial production and economic change to attain competitive advantage and sustainable economic and social growth. This indicates the necessity of a close cooperation between the government and industrial companies. There will be many differences in responding to these dramatic changes of industrial and economic change among countries and industrial companies.

3. Korean approach to Industry 4.0: the 4th Industrial Revolution

3.1. Total mobilization system and its divergent regional disparity

Many countries prefer the concept of Industry 4.0 to the 4th Industrial Revolution. Germany uses Industry 4.0 and it is one of the major promotion areas of the 'High-tech-Strategie', which is the representative S&T policy programme (BMBF 2016b). However, in Korea, the concept of the 4th Industrial Revolution has been more widely used, especially because the current Korean government has adopted the preparation for the 4th Industrial Revolution as one of the major national policy agendas.

During the six decades of its industrialization, Korea has transformed from a poor country to a developed one. Since the end of the 1960s, Korea has formulated and implemented a total S&T mobilization system, which can be termed nowadays as national innovation system (NIS) (STEPI 1992; OECD 2009; Kim 1997; Chung 2003, 2016). During those decades, Korea increased its national R&D investment and manpower dramatically (Chung 2003, 2016, 2019). As a result, nowadays Korea is recognized as one of the most R&D intensive countries in the world. In 2015, Korea was ranked as the most R&D intensive country in the world, by showing 4.22% of total R&D investment as a share of GDP (OECD 2018). However, the Korean industrialization has been concentrated mainly in southeastern coastal regions, leading to these regions being in a better position of increasing their R&D and innovation capabilities and preparing for the 4th Industrial Revolution.

Based on the total mobilization system, Korean S&T and national economy have developed dramatically. Korea could produce and export lots of high-tech products for international markets. However, there have been regional concentrations of industrial development such as shipbuilding (Ulsan[2]), petrochemicals (Ulsan, Jeonnam), machinery (Gyeongnam), automobile (Ulsan), semiconductor (Gyeonggi) and so on. In particular, Korea has been very competitive in many ICT areas, so that it can better prepare for the 4th Industrial Revolution or digitalization. It is interesting that ICT sectors are concentrated in Gyeonggi, where most industrial sectors except heavy and chemical industrial sectors tend to be located. These regional concentrations could generate the problems of differential potential for the 4th Industrial Revolution across regions.

It seems that Korea will have difficulties to prepare for the 4th Industrial Revolution and the Industry 4.0 because most R&D and technological capabilities are concentrated in Seoul and its outskirts (Gyeonggi, and Incheon). As of 2019, about 65% of Korean R&D actors, about 79% of total R&D expenditures, and about 72% of total R&D personnel are concentrated in these regions. In particular, the Gyeonggi Province has a share of 34.47%, 61.46% and 44.77%, respectively, (MSIT & KISTEP 2020; also see Table 4). Most Korean companies have wanted to establish their R&D institutes in these regions because most Korean well-qualified manpower prefer to live and work in these regions. It seems that such a high degree of regional concentration would impede Korea's preparation for the 4th Industrial Revolution.

3.2. Korean approach to Industry 4.0: the 4th Industrial Revolution

3.2.1. Presidential Committee and a concentration of decision-making

Korea recognizes the importance of the 4th Industrial Revolution very seriously. Korea expects that the Revolution will give rise to innovative transformation not only in industries but also in the national system, society and everyday lives (PCFIR 2017). Therefore, the Korean government established the 'Presidential Committee on the Fourth Industrial Revolution' (PCFIR) in October 2017.

The main mission of the Presidential Committee is to inquire and coordinate major relevant activities of Korean ministries to respond to the 4th Industrial Revolution. The major ministers such as the Minister of Science and ICT (MSIT), the Minister of Trade, Industry and Energy (MOTIE) and the Minister of SMEs and Startups (MSS) are members of this Committee. It implies that almost all ministries should implement their own policies to prepare for the 4th Industrial Revolution following the guidance of the PCFIR. However, regional governors are not members of the Committee and there is no systematic way for them to participate in the national efforts for responding to the 4th Industrial Revolution. Figure 2 shows the organizational structure of the PCFIR.

The major tasks of the Presidential Committee comprise four main areas (PCFIR 2017): (1) Early Realization of the 4th Industrial Revolution through the Promotion of 'Intelligent Technology Innovation Projects', (2) Secure Technologies as Growth Engines, (3) Establish Industrial Infrastructure Ecosystem and (4) Responses to Future Social Changes. There are 12 intelligent technology innovation projects and 8 policy-related agendas on going. Therefore, the tasks of the Presidential Committee cover almost all areas of national administration. The Committee initiated the Masterplan for Preparing for the 4th Industrial Revolution ('I-Korea 4.0') in November 2017 and detailed plans in individual projects and agendas have been prepared and implemented by relevant ministries.[3] However, there is no policy agenda for regional preparation for the 4th Industrial Revolution.

Therefore, in preparing for the 4th Industrial Revolution, the S&T-oriented ministries should play an important role. Each ministry plays the main role in implementing its own

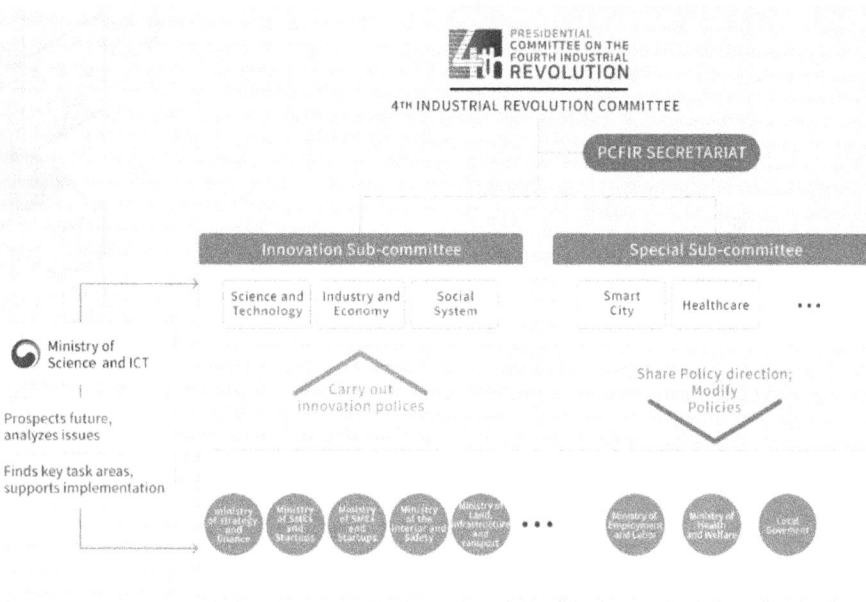

Figure 2. Presidential Committee on the 4th Industrial Revolution (PCFIR). (Source: www.4th-ir.go.kr/home).

action plan to put into practice the Revolution. Under the systematic guidance of the ministries, relevant industrial companies, public research institutes and universities should actively participate in these action plans. In this sense, the total mobilization system has been applied to detailed technologies (for example, smart factory, big data, convergence among relevant technologies and so on) and major industrial sectors (for example, healthcare, agriculture, traffics, environment and do on), which are related to the 4th Industrial Revolution. However, there have been no concrete agendas for regional preparation for the 4th Industrial Revolution in the Presidential Committee, even though these technologies and sectors are essential for regional development.

3.2.2. Korea's potential to respond from regional perspectives
The Korean approach to the 4th Industrial Revolution is very comprehensive and covers almost all aspects of national administration including almost all industrial sectors. In addition, most policy agendas and projects are technology-oriented so that Korea has tried to prepare for the 4th Industrial Revolution through R&D and technological capabilities. Therefore, it would be interesting to investigate whether the policy agendas and projects, formed by the Presidential Committee, can effectively attain the goals across the nation. As Korea is a centralized country, it would be particularly interesting to investigate whether these policy agendas and projects could be implemented and handled at the regional level. In this regard, it would be essential to analyse the R&D and technological capabilities in Korean regions.

In fact, Korea has developed the national economy through industrial clusters since the beginning of its industrialization. In the early 1970s, the Korean government started to nurture heavy and chemical industries and several industrial complexes were established in the south-eastern and south coast of South Korea. These policies have been successful but caused the problems of uneven economic and social development among Korean regions. Since Korea is a traditionally centralized country, industrial activities have been concentrated in Seoul, the capital city of South Korea and Gyeonggi Province, which surrounds Seoul. Industrial companies and their R&D units have been concentrated in both regions and the other industrialized regions discussed above.

Table 2 indicates the regional distribution of R&D performing units in Korea by actor group. The R&D performing units of Korean private companies, i.e. private research institutes, are heavily concentrated in Gyeonggi Province (34.90% of total private research institutes) and Seoul Metropolitan City (24.29%). Gyeonggi Province has a geographical advantage, due to its proximity to Seoul. Regardless of industrial sectors, many SMEs are operating in this region, and thus their research institutes have been established and operating in these regions. Most private research institutes in Seoul do not belong to

Table 2. R&D performing units by actor group and region (2018).

Region		Public Res. Institutes (GRIs)		Univ. and Colleges		Business enterprises (Private companies)		Total	
		Number	%	Number	%	Number	%	Number	%
Metropolitan cities	Seoul	128 (22)	20.92 (13.17)	88	20.71	12,319 (12,312)	24.28 (24.29)	12,535	24.21
	Busan	35 (12)	5.72 (7.19)	24	5.65	2241 (2241)	4.42 (4.42)	2300	4.44
	Daegu	29 (11)	4.74 (1.92)	16	3.76	1,939 (1937)	3.83 (3.82)	1984	3.83
	Incheon	19 (2)	3.10 (1.20)	15	3.63	2970 (2966)	5.85 (5.85)	3004	5.80
	Gwangju	21 (9)	3.43 (5.39)	16	3.76	870 (869)	1.71 (1.71)	907	1.75
	Daejeon	41 (25)	6.70 (14.97)	21	4.94	1576 (1564)	3.11 (3.09)	1638	3.16
	Ulsan	14 (6)	2.29 (3.59)	5	1.18	625 (624)	1.23 (1.23)	644	1.24
	Sejong	17 (15)	2.78 (8.98)	5	1.18	169 (169)	0.33 (0.33)	191	0.37
Provinces	Gyeonggi	67 (11)	10.95 (6.59)	79	18.59	17,697 (17,690)	34.88 (34.90)	17,843	34.47
	Gangwon	32 (7)	5.23 (4.19)	19	4.47	620 (618)	1.22 (1.22)	671	1.30
	Chungbuk	28 (8)	4.58 (4.79)	18	4.24	1640 (1639)	3.23 (3.23)	1686	3.26
	Chungnam	22 (3)	3.59 (1.80)	28	6.59	1865 (1863)	3.68 (3.68)	1915	3.70
	Jeonbuk	38 (11)	6.21 (6.59)	20	4.71	1019 (1017)	2.01 (2.01)	1077	2.08
	Jeonnam	30 (5)	4.90 (2.99)	15	3.53	723 (720)	1.43 (1.42)	768	1.48
	Gyeongbuk	41 (7)	6.70 (4.19)	24	5.65	1801 (1799)	3.55 (3.55)	1866	3.60
	Gyeongnam	34 (10)	5.56 (5.99)	26	6.12	2493 (2490)	4.91 (4.91)	2553	4.93
	Jeju	16 (3)	2.61 (1.80)	6	1.41	165 (164)	0.33 (0.32)	187	0.36
Total		612 (167)	100 (100)	425	100	50,732 (50,682)	100 (100)	51,769	100

Source: MSIT & KISTEP (2019).

heavy and chemical industries but belong to ICTs and other industries which do not need a lot of space and facilities. Metropolitan cities such as Incheon (5.85%), Busan (4.42%) and Daegu (3.83%) have many private research institutes, especially compared to most other provinces. Among provinces, Gyeongnam (4.91%), Gyeongbuk (3.55%), Chungnam (3.68%) and Chungbuk (3.23%) have relatively many private research institutes compared to other provinces. This implies that about half of Korean regions seem to have a certain level of R&D and technological capabilities, even though most Korean industrial R&D capabilities are concentrated in Gyeonggi and Seoul.

Compared to the R&D performing units of industrial companies, public research institutes and universities and colleges are relatively fairly distributed. Public research institutes, especially government-sponsored research institutes (GRIs), have played an important role in the development of Korea's S&T and economy. Before the 2000s they were heavily concentrated in Seoul and Daejeon. The Korean government has completed the establishment of Daedeok Science Town in Daejeon, where most of the Korean GRIs were established. Several big enterprises also placed their research institutes in this town. However, recognizing the importance of the regional distribution of R&D and technological capabilities, the Korean government has implemented regional innovation policies since the beginning of the 2000s. Korean GRIs have established their branch institutes in regions, which have become the focal point of regional innovation systems (Chung 2016; MSIT 2019). Korean universities and colleges have been also regionally well distributed, compared to other actor groups. Historically the Korean government established a public university per region (province) at the beginning of the industrialization and also established S&T-specific universities in Korean regions to further strengthen regional R&D and innovation capabilities, i.e. Gwangju in 1993, Ulsan in 2009 and Gyeongbuk in 2003.

It seems that many Korean regions implement relatively good regional innovation systems, so that Korean industry-GRIs-universities linkages have been strengthened in regions, even though more developed regions such as Gyeonggi, Gyeongbuk, Gyeongnam and Ulsan are in a better position for implementing regional innovation systems. However, Korea should take the regional differences of R&D and innovation capabilities into account to effectively prepare for the 4th Industrial Revolution. Heavily industrialized regions have had a big advantage in preparing for the 4th Industrial Revolution. They are much more R&D- and technology-intensive than other regions. For example, these regions have shown a much stronger demand for 'smart factories', which are important elements of the Industry 4.0.

4. A case: regional diffusion of smart factories in Korean SMEs

4.1. Introduction to Korean smart factories

The smart factory is a Korean version of Industry 4.0. It is one of the major agendas of Korea's preparation for the 4th Industrial Revolution. Korean industrial companies also recognized the importance of smart factories in attaining their industrial competitiveness and have made great efforts to adopt them. In 2014, the Korean government has introduced the 'Program for Diffusing and Advancing Smart Factory' for Korean industrial companies (MOTIE 2017). As the government recognized the importance of the 4th

Industrial Revolution, the programme has been extended and a series of sub-pro-
grammes have been implemented.

The programme was launched centrally, although its purpose was to diffuse smart factories in SMEs all over the regions. The programme was started in June 2014 by the Ministry of Trade, Industry and Energy (MOTIE), but in July 2017 was transferred to the
Ministry of SMEs and Startups (MSS), an extended ministry from the Small and
Medium Business Administration (SMBA). In this programme, the government, big
enterprises and SMEs have been cooperating very closely. From the governmental
sector, Ministry of SMEs and Startups (MSS), Ministry of Trade, Industry and Energy
(MOTIE) and Ministry of Science and ICT (MSIT) are actively participating, and
about 50 Korean big enterprises, for example Samsung Electronics, Samsung Display,
POSCO and Hyundai Automobiles are very active in the programme. The budget of
the programme is composed of the governmental budget and the Co-existent Fund
between Big and Small Enterprises that is supplied by big enterprises. When a SME is
selected and accepted to the programme, it must invest its own money as a matching
fund. As this programme is multi-ministerial, public-private collaborative programme,
there was a strong need of coordination and management of the programme, so the government established the Korea Smart Manufacturing Office (KOSMO) in 2015.

This programme classifies smart factories into four levels based on the technologies
involved: Basic Level, Middle 1 Level, Middle 2 Level and Advanced Level (PCFIR
2018). When applying for the programme, SMEs can select the level, considering their
technological and financial situations. In general, the programme covers up to half of
the costs of establishing a smart factory. As for the project level of smart factories,
there are two kinds of project: new establishment project and advancement project
(MSS 2020).

In general, SMEs, which have interest in applying for the programme, should build a
consortium with a supplying company, which provides the services of establishing smart
factories for the SMEs. There are two kinds of promotion in this programme: government-led and co-existence-type promotion. Under the 'government-led' type, which is
often called general type, a SME should build a consortium with a supplying company
and the government supports the full amount of programme support, which is less
than half of the total costs of establishing a smart factory. The other amount should
be covered by SMEs seeking to establish a smart factory. Under the 'co-existent' type,
a SME should build a consortium with a big company and the government supports a
maximum of 30% of the total costs of establishing a smart factory, the big company
covers more than 40% and the SME covers less than 30% of the total costs. In this
type, the participating big enterprise and SMEs tend to have a parent-vendor relationship
(MSS 2020).

4.2. Diffusion of smart factories based on technological levels

This programme started in June 2014 and aimed at cumulatively establishing 20,000
smart factories in Korean manufacturing industries by the end of 2022, but the target
number has been extended to 30,000 smart factories in 2018 due to the strong
demand of Korean SMEs for smart factories. According to the survey of KOSMO,
there has been a strong increase in smart factory establishment, cumulatively from 277

factories in 2014 through 2,800 factories in 2016–7,903 factories in 2018 (KOSMO 2019), 28.5 times larger than 4 years back.

Looking at the level of establishment of smart factories (Table 3), 76.4% of Korean SMEs established the Basic Level that concentrated on the digitalization of production information and the management of production track record. 21.5% of SMEs established the Middle 1 Level that enables real-time gathering and analysis of production information. There is no SMEs that adopted smart factories of Advanced Level, and only 2.1% of Korean SMEs adopted the Middle 2 Level that could control the production process through an efficient system. It implies that Korean SMEs are at the beginning of adopting smart factories. Therefore, Korean SMEs should establish smart factories of more advanced level to attain competitiveness.

4.3. Diffusion of smart factories in Korean regions

In Korea, most smart factories were established in industrialized regions. According to the survey of KOSMO (see Table 4), as of the end of 2017, Gyeonggi (1,251) established the largest number of smart factories, followed by Gyeongbuk (736), Gyeongnam and Daegu (479). Gyeonggi, the outskirt province of Seoul, has the strongest demand for smart factories among Korean regions because it is where most Korean SMEs are located. The other three regions are in the southeastern part of South Korea and have been traditionally industrialized. It is interesting that Seoul (83 factories) and Busan (275 factories) did not have enough smart factories. These are the first and second largest cities in Korea and did not have enough manufacturing facilities. In addition, there was no sufficient diffusion of smart factories in the southwestern part of South Korea (Gwangju, Jeonnam and Jeonbuk) and the north-eastern part (Gangwon). They are less developed and mainly agricultural regions. As a result, we can classify the regional diffusion of Korean smart factories into three categories: actively adopting regions (4 regions), moderately adopting regions (5 regions) and inactively adopting regions (8 regions). It seems that different categories need different policy measures for the effective diffusion of Korean smart factories.

According to KOSMO (2019), most Korean smart factories were established in sectors such as transportation parts and equipment (27.4%), machinery (27.3%), electronics and electrical parts (15.2%) and petrochemicals (14.3%). Therefore, smart factories were diffused based on the regional location of Korean industrial sectors. For example, industrial companies in machinery are concentrated in Gyeongnam and Gyeongbuk, petrochemical companies in Gyeongnam and Ulsan, electronics and electrical parts

Table 3. Level of Korean smart factories established.

Level	Characteristics	Share of SMEs established
Basic Level	Digitalization of production information and management of production track record	76.4%
Middle 1 Level	Real-time gathering and analysis of production information	21.5%
Middle 2 Level	Control of production process through system	2.1%
Advanced Level	Customized flexible manufacturing and intelligent factory	–

Source: PCFIR (2018).

Table 4. Diffusion of smart factories and industrial R&D intensity of Korean regions.

		Distribution of smart factories in 2017	R&D expenditures in 2018	
Classification	Regions	Number of smart factories established	Amount (Bill. Won)	%
Actively adopting regions	Gyeonggi	1251	41,749	61.46
	Gyeongbuk	736	2351	3.46
	Gyeongnam	519	2081	3.06
	Daegu[a]	479	779	1.15
Moderately adopting regions	Chungnam	320	2097	3.09
	Incheon[a]	301	2232	3.29
	Gwangju[a]	283	393	0.58
	Busan[a]	275	733	1.08
	Chungbuk	249	1119	1.65
Inactively adopting regions	Jeonbuk	127	450	0.66
	Ulsan[a]	121	855	1.26
	Jeonnam	99	338	0.50
	Seoul[a]	83	9746	14.35
	Gangwon	72	196	0.29
	Daejeon[a]	56	2585	3.81
	Sejong[a]	24	181	0.27
	Jeju	8	41	0.06
Total		5003	67,926	100

[a]Metropolitan cities.
Source: Classified from KOSMO (2019); MSIT & KISTEP (2019).

companies in Gyeonggi. Industrial companies in transportation parts are rather region-ally well diffused because Korean automobile factories are regionally distributed. Korean companies in almost every sector are concentrated in Gyeonggi province because it sur-rounds the capital city, Seoul. Therefore, the national strategy in the 1970s and 1980s for establishing heavy and chemical industrial clusters along the southeastern coast led to a relatively stronger demand for smart factories in these regions.

To effectively diffuse smart factories in Korea, particularly in Korean regions, techno-logical capabilities of demander SMEs and suppliers of smart factories should be enhanced. Technological capabilities are essential for demander companies to establish higher level of smart factories. They should have a certain level of 'absorptive capacity' (Cohen and Levinthal 1990) to adopt and exploit higher-level smart factories. As dis-cussed above, smart factories are not evenly diffused in Korean regions. To solve this problem and to motivate SMEs to adopt smart factories of more advanced level, regional distribution of technological capabilities should be taken into consideration. SMEs with a higher level of R&D capabilities would adopt smart factories of more advanced level.

Table 4 shows also that there are the strong differences in the industrial R&D and tech-nological capabilities of Korean regions. Gyeonggi has the strongest industrial R&D capabilities, substantially larger than those of other regions, accounting for 61% of total industrial R&D expenditures. Seoul is the second most R&D-intensive, accounting for 14% of total industrial R&D expenditures. Except Gyeonggi and Seoul, we can see pat-terns similar to the number of R&D performing actors (See Table 2).

To implement more effective policies to diffuse smart factories, the level of diffusion and R&D intensity should be taken into consideration. Figure 3 shows the typology of Korean regions and we can see the challenges that government policies have to manage. Different levels of R&D capabilities imply that Korean regions have different challenges in the programme for diffusing smart factories.

First, Seoul and Gyeonggi are not in need of governmental support. Seoul is a metro-politan service-oriented city and has only a few industrial factories. Gyeonggi has the

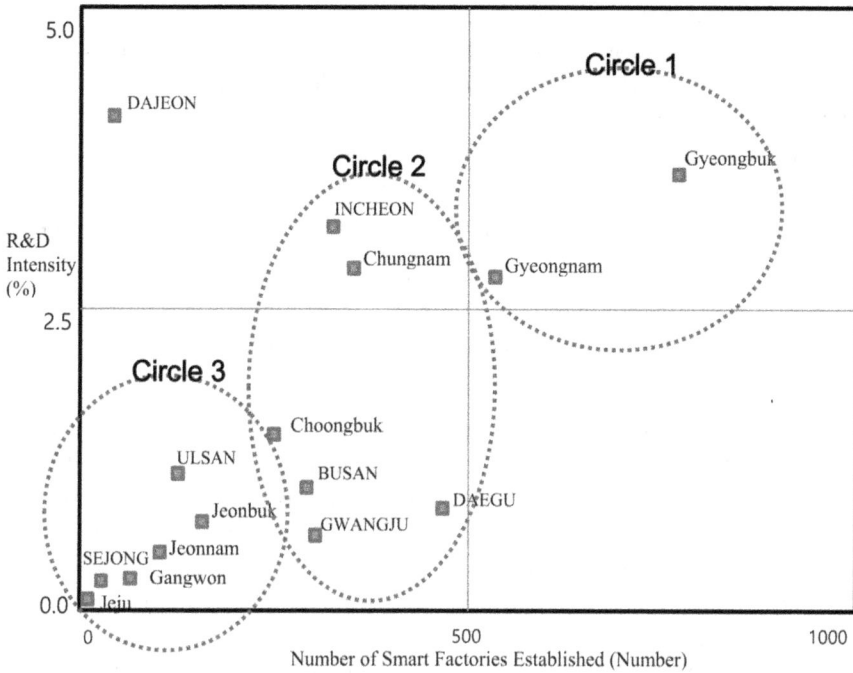

Figure 3. Typology of Korean regions in the diffusion of smart factories. *Regions in capital letters are metropolitan cities.

largest number of Korean SMEs and a larger number of smart factories than other regions. It seems that SMEs in both regions should establish smart factories with their own budgets and interests.

Second, SMEs in the actively adopting regions (Circle 1), especially Gyeongbuk and Gyeongnam, should be supported by the 'co-existent' type of the programme, because they are relatively familiar with smart factories since many other SMEs have already adopted smart factories. In addition, SMEs that have already adopted smart factories should be motivated by the governmental programme to adopt factories of a more advanced level. SMEs in these regions are more R&D-intensive than those in other regions, and thus they are in a better position to adopt more technology-intensive smart factories.

Third, the moderately adopting regions (Circle 2), especially Busan, Incheon, Chungnam and Chungbuk, and an actively adopting region, Daegu should be supported by a dual approach. SMEs in these regions should be supported by either 'government-led' or 'co-existent' promotion of the programme. SMEs that have not yet adopted smart factories should be supported for their first adoption and SMEs that have already adopted should be supported for their re-adoption of more-advanced-level smart factories. Regarding the diffusion of more-advanced-level smart factories, Incheon and Chungnam are in a good position since they have relatively very strong R&D intensity. Other regions, such as Chungbuk, Busan and Daegu also have relatively strong R&D intensity. This issue should also be considered for the implementation of the governmental programme.

Finally, others are the inactively adopting regions (Circle 3) and are not R&D intensive. They are less industrialized agricultural regions, particularly Jeonbuk, Jeonnam, Gangwon and Jeju. As they show a very low level of diffusion, SMEs in these regions should be motivated and supported to adopt smart factories for the first time, especially by the 'government-led' promotion. It should also be taken into account that SMEs in these regions have a very low-level of R&D intensity. Therefore, they should adopt lower-level smart factories through the active support by the governmental programme.

5. Conclusions for Industry 4.0 and uneven regional development

The Korean approach to Industry 4.0 is to prepare for the 4th Industrial Revolution, which is characterized by a total mobilization system. The Korean government established the 'Presidential Committee on the 4th Industrial Revolution (PCFIR)' in 2017, which covers almost every aspect of national administration. All relevant ministries have implemented their policy agendas respectively and almost all relevant actors are mobilized in these agendas. Many policy agendas of PCFIR are technology-oriented so that Korean national and regional R&D and technological potentials have played an important role in the Korean approach. However, the Korean approach has limitations in its implementation, particularly due to the uneven distribution of R&D and technological capabilities across regions. This issue might aggravate the uneven economic development of regions in South Korea.

One of the representative projects of PCFIR is the diffusion of smart factories in Korean SMEs. This project has been handled by the Ministry of SMEs and Startups (MSS), and many big enterprises, SMEs, and public organizations have been cooperating very closely. Based on this project, about 8000 smart factories have been established by Korean SMEs as of the end of 2018 and the Korean government has a goal of diffusing 30,000 smart factories cumulatively by 2022. However, the diffusion of smart factories would entail difficulty in attaining their policy goals, because Korean R&D and innovation potentials are concentrated in a few regions, especially in Gyeonggi as well as Gyeongbuk and Gyeongnam.

As for the Korean preparation for the 4th Industrial Revolution, we can identify a few important challenges for Korea, especially from regional perspectives. First, Korea needs to take regional diversity into consideration very seriously. There have been big differences in R&D and technological capabilities across Korean regions, which are essential elements of implementing the Industry 4.0. This is especially important since most of PCFIR's agendas are technology-oriented.

Second, Korean regional governments should be members of PCFIR and actively participate in the planning and implementation of its agendas. They should actively cooperate with ministries and provide essential input for the Presidential Committee from regional perspectives. Even though they have tried to prepare for the 4th Industrial Revolution based on their own interests, they have had no systematic way of connecting with national efforts.

Second, Korea should strengthen R&D and technological capabilities in many regions, as Korean R&D capabilities are concentrated in Seoul and Gyeonggi regions as well as in the southeastern part of South Korea, Gyeongnam and Gyeongbuk. This regional discrepancy may impede Korea's preparation for the 4th Industrial Revolution. The number of regional R&D performing actors should be increased and their R&D intensity should be strengthened. Korean SMEs in regions should enhance their international competitiveness in terms of actively adopting and advancing smart factories. Regional governments, which have been inactive in promoting R&D activities and preparing for the 4th Industrial Revolution, should also increase their R&D investment, since the R&D investment of Korean regional governments is very low, compared to that of regional governments in advanced countries such as Germany.

Third, particularly regarding the diffusion of smart factories, Korea should implement a series of policy measures to generate synergy effects between SMEs' adoption of smart factories and their R&D capabilities, especially in less developed regions. Based on the diffusion level of smart factories and the technological capabilities of different Korean regions, different policies should be implemented to generate synergy effects and to effectively enhance the competitiveness of SMEs in regions. The governmental promotion should focus on SMEs with weak financial and technological capabilities, particularly in less industrialized regions (Chung, Jeon, and Hwang 2016).

Finally, the Korean approach to the 4th Industrial Revolution and the diffusion of smart factories should be pursued continuously since it is essential for the long-term development of Korean economy and society. In Korea, there has been a tendency of discontinuous national policy agendas through the change of governmental administration, particularly since the beginning of the 2000s. However, the agenda of diffusing smart factory, which was started by the last administration, have been succeeded and extended by the current administration, and allows us to expect fruitful results. The preparation for the 4th Industrial Revolution and its detailed policy agendas should continue to be implemented by relevant ministries, regional governments and industrial companies.

Notes

1. For example, the Korean government established a government-sponsored research institute, Electronics and Telecommunication Research Institute (ETRI), in Daedeok in 1976, which is in the center of South Korea. It has made a great contribution to the development of most Korean ICTs, for example, computers, semiconductor, telecommunications, systems engineering and so on. ETRI has concentrated on making smart Korea in the 2010s and leading the 4th Industrial Revolution since the middle of the 2020s (ETRI 2017).
2. Ulsan belonged to Gyeongnam province and, due to its industrial development, it was separated into and become a metropolis in1997.
3. The mission of the Committee is to discuss and coordinate the major related activities of all relevant ministries regarding the 4th Industrial Revolution. For two years since the establishment of the Committee there have been ten official meetings to discuss major issues such as hyper-connected smart network (3rd meeting), smart city (4th meeting), smart factory (5th meeting), AI R&D (6th meeting), data industry (7th meeting) and cloud computing (8th meeting). Most of agendas for the meetings have been prepared by relevant ministries (PCFIR 2018).

Acknowledgement

This paper was supported by Konkuk University in 2019.

Disclosure statement

No potential conflict of interest was reported by the author(s).

References

Bundesministeriun für Bildung und Forschung (BMBF). 2016a. *Bundesbericht Forschung und Innovation 2016*. Berlin: BMBF.

Bundesministeriun für Bildung und Forschung (BMBF). 2016b. *Industrie 4.0: Innovationen für die Produktion von Morgen*. Berlin: BMBF.

Carvalho, N., O. Chaim, E. Cazarini, and M. Gerolamo. 2018. "Manufacturing in the Fourth Industrial Revolution: A Positive Prospect in Sustaining Manufacturing." *Procedia Manufacturing* 21: 671–678. doi:10.1016/j.promfg.2018.02.170

Chung, S. 1996. *Technologiepolitik für neue Produktionstechnologien in Korea und Deutschland: Einfluß der länderspezifischen Rahmenbedingungen auf die Technologiepolitik*. Heidelberg: Physica-Verlag.

Chung, S. 2002. "Building a National Innovation System Through Regional Innovation Systems." *Technovation* 22 (8): 485–491. doi:10.1016/S0166-4972(01)00035-9

Chung, S. 2003. "Innovation in Korea." In *The International Handbook on Innovation*, edited by L. V. Shavina, 890–903. Oxford: Pergamon.

Chung, S. 2016. "Korean Government and Science and Technology Development." In *Routledge Handbook of Politics and Technology*, edited by U. Hilpert, 222–235. London: Routledge.

Chung, S. 2019. "South Korea as a New Player in Global Innovation: Role of a Highly Educated Labour Force's Participation in New Technologies and Industries." In *Diversities of Innovation*, edited by U. Hilpert, 179–209. London: Routledge.

Chung, S., J. Jeon, and J. Hwang. 2016. "Standardization Strategy of Smart Factory for Improving SME's Global Competitiveness." *Journal of Korea Technology Innovation Society* 19 (3): 545–571 (Korean).

Cohen, W., and D. Levinthal. 1990. "Absorptive Capacity: A New Perspective on Learning and Innovation." *Administrative Science Quarterly* 35 (1): 128–152. doi:10.2307/2393553

Electronics and Telecommunications Research Institute (ETRI). 2017. *A 40-Year History of ETRI*. Daejeon: ETRI (Korean).

Freeman, C., and C. Perez. 1988. "Structural Crises of Adjustment: Business Cycles and Investment Behavior." In *Technical Change and Economic Theory*, edited by G. Dosi, 38–66. London: Pinter.

Gold, B. 1989. "Harnessing the Capabilities of CIM: The Critical Role of Senior Management." *Research Policy* 18: 173–181. doi:10.1016/0048-7333(89)90005-X

Hwang, D., Y. Cheong, and S. Chung. 2019. "The Evolution of Innovation Cluster: Focusing on the Daedeok Innopolis." *Journal of Korea Technology Innovation Society* 21 (4): 1207–1236 (Korean).

Kim, L. 1997. *Imitation to Innovation: The Dynamics of Korea's Technological Learning*. Boston, MA: Harvard Business School Press.

Kondratieff, N. 1935. "The Long Waves in Economic Life." *The Review of Economic Statistics* 17 (6): 105–115. doi:10.2307/1928486

Korea Smart Manufacturing Office (KOSMO). 2019. *Internal Data*. Daejeon: KOSMO (Korean).

Lay, G. 1993. "Government Support of Computer Integrated Manufacturing in Germany: The First Results of an Impact Analysis." *Technovation* 13: 283–297. doi:10.1016/0166-4972(93)90002-D

Ministry of Science and ICT (MSIT). 2019. *Regional S&T Yearbook 2018*. Seoul: MSIT (Korean).

Ministry of Science and ICT (MSIT)/Korean Institute of S&T Evaluation and Planning (KISTEP). 2019, 2020. *<Survey of R&D in Korea*. Seoul, December (Korean).

Ministry of SMEs and Start-ups (MSS). 2020. *Public Announcement of the Program for Suppling and Diffusing of Smart Factories in 2020*. Daejeon: MSS (Korean).

Ministry of Trade, Industry and Energy (MOTIE). 2017. *Vision of Smart Manufacturing Innovation 2025*. Seoul: MOTIE (Korean).

Neugebauer, R., S. Hippmann, M. Leis, and M. Landjerr. 2016. Industrie 4.0 – From the Perspective of Applied Research, The 49th CIRP Conference on Manufacturing System, *Procedia CIRP, 2-7* (http://creativecommons.org/licenses/by-nc-nd/4.0/)

OECD. 2009. *OECD Reviews of Innovation Policy*. Paris, Korea: OECD.

OECD. 2018. *Main Science & Technology Indicators 2018-1*. Paris: OECD.

Perez, C. 2002. *Technological Revolutions and Financial Capital: The Dynamics of Bubbles and Golden Ages*. Cheltenham: Edward Elgar.

Presidential Committee on the Fourth Industrial Revolution (PCFIR). 2017. People-Centered "Plan for the Fourth Industrial Revolution" to Promote Innovative Growth (I-Korea 4.0). Seoul.

Presidential Committee on the Fourth Industrial Revolution (PCFIR) and All Related Ministries. 2018. Strategies for Diffusing and Enhancing Smart Factories. The Fifth Resolution Agenda of PCFIR, March, Seoul (Korean).

Presidential Committee on the Fourth Industrial Revolution (PCFIR) and All Related Ministries. 2018. The Progress Status of the Reviewed and Resolved Agendas of PCFIR. July, Seoul (Korean).

Rifkin, J. 2011. *The Third Industrial Revolution: How Lateral Power is Transforming Energy, the Economy, and the World*. New York: Palgrave Macmillan.

Schumpeter, J. 1911. *Theorie der wirtschaftlichen Entwicklung*. Leipzig: Auflage. 1

Schumpeter, J. 1939. *Business Cycles: A Theoretical, Historical, and Statistical Analysis of the Capitalist Process*. New York: McGraw-Hill.

Schwab, K. 2016. *The Fourth Industrial Revolution*. New York: Crown Business.

Science and Technology Policy Institute (STEPI). 1992. *National Innovation System in Korea*. Seoul: STEPI. (Korean). www.4th-ir.go.kr/ www.smart-factory.kr/

Industry 4.0/Digitalization and networks of innovation in the North American regional context

Paul M.A. Baker, Helaina Gaspard and Jerry A. Zhu

ABSTRACT

The advancement of industrial, innovation-related economic policies such as Industry 4.0 and the advanced digitalization of production play an increasingly important role in fulfilling economic objectives in both Canada and the United States. There are a variety of ways in which such industrial-related policy approaches can be developed and implemented. Varying aspects of industrial and economic innovation often occur within a regional context, which can change policy is developed and implemented, dramatically and with little warning. This paper applies a case-based approach to examine enabling and constraining factors of regional innovation policy in two cases - Ontario, Canada and Massachusetts, US. Moving beyond a linear conception of regional innovation, this research explores how policies and modalities for collaboration can facilitate Industry 4.0 and related innovation ecosystems. Our analysis suggests that regional innovation impact is influenced through four principal factors: industrial clusters; context; collaborative synergies; and network intermediaries. Additional research could focus on an expanded case examination of the relationship between top-down policy approaches and the operation of regional innovation ecosystems coupled with bottom-up market- and stakeholderdriven analytic approaches.

Introduction

The advancement of industrial and innovation-related economic policies, such as the emerging Industry 4.0 (or 'smart manufacturing') and the advanced digitalization of production, plays an increasingly important role in fulfilling economic objectives in both Canada and the United States. There are a variety of ways in which industrial-related policy approaches can be developed and implemented. Policy may focus on approaches that augment the production side, resulting in change or impact on manufacturing or production processes. However policy is developed and implemented, different aspects of industrial and economic innovation often occur within a regional context, which can change dramatically and with little warning. This became especially apparent during the development of this paper, which occurred during the COVID-19 pandemic in early 2020, underscoring the observation that all types of longer-term policy are impacted by intervening events.

Consider the Industry 4.0 model, which can be thought of as the integration of cyber-infrastructures in production and logistics as well as the application of the Industrial Internet of Things (IIoT) to create smart manufacturing capacity (Mittal, et al. 2019). Here, Intelligent capacity is developed in horizontally and vertically integrated production systems that lead to consistent engineering throughout the process (Reynolds & Uygun, 2018). Many aspects of the Industry 4.0 approach apply to other types of economic activity including services. An ancillary concept, 'digitalization', can be seen in manufacturing and associated services in the target cases, but more generally in the operation of platform based production in both a classical and 'post-manufacturing' sense.

Many of the information-based advances driving innovation are associated with geographically centred nodes of technical development and creativity. While the most iconic of these is Silicon Valley, other recognized North American nodes of innovation include Seattle (aviation and software), Boston (software, fintech, robotics) and New York (entertainment, fintech, media, ecommerce), and emerging nodes such as Nashville (healthcare IT, biotech/pharma, entertainment), Salt Lake City (edtech, information and telecom technology), Denver (aviation/aerospace, telecommunications, energy), Austin (technology, entertainment), Miami (media, consumer technology), Raleigh-Durham (Biotech/Pharma) and Atlanta (logistics, telecommunications, IT) in the United States (Muro and Liu, 2017); and Vancouver (digital technologies, life sciences, entertainment/media), Toronto (fintech, AI, augmented reality/virtual reality (AR/VR), cybersecurity) and Montreal (software/IT and pharma) in Canada (Muro, et al., 2018).

As the examination of these areas becomes more nuanced, the manifestation of these digital innovations has begun to expand from the underlying digital technologies to how these are deployed and diffused to adjacent and ancillary fields. Preliminary work suggests that an examination of the role of intermediaries would be a useful addition to the research focus (Clayton, Feldman, & Lowe, 2018; Coenen et al., 2017; De Silva, Howells, & Meyer, 2018; Kivimaa, et al., 2017; Gaspard, Baker, & Zhu, 2019).

In this chapter, we adopt a case-based approach to examine regional innovation, and enabling and constraining factors of regional innovation policy by exploring:

1. Industry network response to digitalization by connecting or building from pre-existing foundations (regional knowledge clusters)
2. Systemic change as a consequence of the emergence of new or adapted institutions/structures (e.g. rules, laws, practices) (context)
3. Network interaction between actors (i.e. organizations, people) engaged in regional innovation (collaborative synergies)
4. Intermediaries' activities and influence on regional networks of innovation through top-down or bottom-up action (discernable role of intermediaries).

We apply this framework to two regional cases – Ontario, Canada and Massachusetts, US – selected as geographical representatives of regions recognized as having leadership in software and digital platform development, as well as networks of science and technology innovation. Adopting a comparative US–Canada approach, the key research objective is to identify social, economic and policy factors associated with the (1) emergence of regions and metropolises as innovation nodes/centres of expertise and (2) the impact of related industrial sectors, in this case software development. Moving beyond a linear conception of regional innovation, this research is designed to explore and highlight how policies and modalities for collaboration 'open' and/or encourage Industry 4.0 and related innovation ecosystems.

Our analysis suggests that regional innovation impact is influenced through four principal factors: industrial clusters; context; collaborative synergies; and network intermediaries. While regional networks of innovation may be fostered through the deliberate collaboration of sectors and actors (top-down), other times it occurs organically, from the bottom-up. In the case of Massachusetts, its regional innovation has been developed around its universities. In Ontario, Toronto's industry focuses on the financial sector. Analysis of the experiences of Massachusetts and Ontario reveal many of the typical drivers of regional innovation, as well as offering contributions to the facilitation of policy and technological innovations.

Regional structures, industry 4.0/Digitalization and innovation networks

In North America, a number of regions have become centres for scientific research and development and technological innovation, a critical component of Industry 4.0 and smart manufacturing (Hilpert, 2019). Industries deeply invested in technology and scientific research and development play a key role in stimulating regional economic opportunities. These relationships are often conceptualized as a regional innovation ecosystem (see, for instance, Hilpert & Lawton Smith, 2013; Frølund, Murray & Riedel, 2017; O'Gorman & Donnelly, 2019), given the interconnectedness and interdependency of the actors. Studies tend to cluster variously by a focus on industry, the role of actors and organizations, as well as by the influence of policy instruments. In the case of industry, the focus is on regional areas of expertise and the strategic behaviour and characteristics of firms (Chang & Hughes, 2012; Ranasinghe, 2017; Wells & Hungerford, 2011). The relationships between the actors, as well as governments are also considered (Frølund, Murray & Riedel, 2017; Tiffin & Kunc, 2011). Analysis of policy instruments tends to focus on the nature of government interventions, the impact of regulations, and the influence of intergovernmental arrangements on the establishment (or stifling), development and growth (or quelling) of sectors (Clark, 2014; Creutzberg, 2011).

A review of pertinent studies revealed at least three gaps in the literature. First, a more nuanced understanding of the flow of knowledge and intangible assets within regional networks is needed (Barrutia et al., 2014; Kramer et al., 2011). Second, measuring the meaningfulness of relationships in regional networks (Tiffin & Kunic, 2011; Pugh, et al., 2018), would help to promote the targeted development of partnerships in emerging networks that may be more likely to yield results. Third, a more robust understanding of the role of intermediaries in regional innovation can support the development and growth of networks of innovation (Wells & Hungerford, 2011).

Approach (Method) and considerations

In this paper, we explore Industry 4.0 and emerging networks of innovation in the target cases to develop generalization of processes and variables characteristic of these regions.

We adopt a case-based approach to explore drivers of regionally based innovation in the software sectors of two sub-national jurisdictions, Massachusetts and Ontario Selecting one Canadian and one US jurisdiction serves as a control to identify whether variables are unique to innovative sectors or moderated by their regional contexts. Both the US and Canada have made commitments to supporting research and development in advanced manufacturing. In 2014, the US government proposed $2.2 billion in advanced manufacturing research and development (President's Council of Advisors on Science and Technology, 2014; Subcommittee on Advanced Manufacturing Committee on Technology of the National Science & Technology Council, 2018). The World Economic Forum's 2018 Future of Production assessment, evaluated production structures in countries around the world, noting that the US was generally well situated to take advantage of Industry 4.0, even though implementation is spotty. In Canada, advanced manufacturing was selected as one of the six sectors the Canadian government targeted for growth through its Economic Strategy Tables (ISED, 2018).

The term 'Digitalization', alternatively, Smart Industry or Smart Manufacturing (Davis et al., 2012), broadens our understanding of Industry 4.0 to include production sectors beyond manufacturing. The software industry, for instance, integrates core principles of Industry 4.0, such as digital translation and cyber-physical systems, without its original connection to manufacturing. Using the principles highlighted by Industry 4.0 as a point of departure allows analysis of networks of innovation to identify and unpack their contributing influences. Analysis of two different regional experiences with the software industry suggests an opportunity to extend the understanding and applications of the contributing elements to these networks of innovation. Through two case studies, we assess the interplay of network innovative practices through the lens of Industry 4.0. Variables for analysis cover considerations of institutional context (e.g. legal-political arrangement, laws, taxation systems), actors (e.g. industry, industry associations, universities) and their inter-activity (e.g. co-development processes, work-integrated learning programs). We explore this series of factors to better explain what drives advancement

Table 1. Population count (2018), education attainment (2018) and the number of intermediaries (2019) in Canada, United States, Ontario and Massachusetts, 2018. Population and education attainment data retrieved from the United States Census Bureau (Table ID S1501 and DP05) and Statistics Canada (Tables 17-10-005-01 and 37-10-0130-01). Intermediary data retrieved from Associations Canada and Gale Publishing, National Center for Education Statistics, Ontario Ministry of Training, Colleges, and Universities.

VARIABLE	CANADA	UNITED STATES	ONTARIO	MASSACHUSETTS
POPULATION (MILLION)	37.6	327.2	14.6	6.9
% OF POPULATION WITH A COLLEGE DEGREE OR HIGHER*	35	33	32	45
TOTAL ASSOCIATIONS	10,828	24,081	7135	740
BUSINESS/FINANCE ASSOCIATIONS	24.79	3521	542	72
ENGINEER ASSOCIATIONS	17.69	1554	595	50
MANUFACTURING ASSOCIATIONS	332	49	168	1
LABOUR ASSOCIATIONS	561	164	147	4
HIGHER ED INSTITUTIONS	96	4298	22	92
OTHER POSTSEC	184	2723	24	117

*Note: For Canada and Ontario, the % of population only includes those between the ages of 25 and 64. For the United States and Massachusetts, the % of population only includes those above the age of 25.

and innovation in an industry predicated on digital and cyber-physical processes characteristic or Industry 4.0 and advanced digitalization of production.

Results and analysis

Industry 4.0 emphasizes the disruptiveness of technology and its impact on the processes
and people involved in the manufacturing sector (Schwab, 2016). While helpful to the-
orize large-scale transformations of manufacturing-based industries, Industry 4.0 as a
sole focus leaves a gap in understanding of how networks of innovation in non-manufac-
turing production sectors are nurtured and changed. With a focus on manufacturing and
advances driven by technology, automation and enhanced analytics, Industry 4.0 is less
useful for understanding service and non-hardware related networks of innovation. To
an extent, this gap is covered by an alternative concept of Digital Industrial Transformation. Aside from Industry 4.0, there are entire highly innovative industrial sectors that
do not manufacture, per se, but are critical for the operation of Industry 4.0: Internet of
Things (IoT): additive manufacturing, digitalization and integration of data and
workflows, remote monitoring, multi-disciplinary engineering and automation of controls through machine learning and predictive analytics. Beyond the role of technology,
how do intermediaries such as lobby or pressure groups, industry associations, etc. and
institutional considerations such as constitutional divisions of power, taxation laws, etc.
shape the operation of service and extractive sectors in different regions?

Table 1 summarizes some of the intermediaries that exist in Ontario and Massachusetts, as well as their respective countries. The table includes population nationally and by
state and province, the educational attainment level, and the number and type of intermediaries present, as well as their primary (headquarters) location. In Ontario, the information and communication technology (ICT)[1] industry has grown in prominence, while
the province's manufacturing sector has shown signs of weakening. Between 2007 and
2017, the value of the ICT industry in Ontario grew 24%, at an average rate of 2.2%
per year, and employment in the industry grew 14%, outpacing the growth of the
Ontario economy (see Table 2) (Statistics Canada, Table 36-10-0402-02; Statistics
Canada, Table 14-10-0202-01). This is mainly driven by an increase in the activities of
the ICT services sector, in contrast to the ICT manufacturing sector, of which the
output from the former represents over 95% of Ontario's ICT industry. Telecommunications and computer systems design (e.g. coders and software developers), two ICT service

Table 2. Growth rate of industries in Ontario from 2007 to 2017. Statistics Canada, table 36-10-0402-
01.

Ontario industry growth (2007–2017)					
Industry	Growth	Average growth rate	% of industry share (2007)	% of industry share (2017)	Change in industry share (2007–2017)
All industries	16.3%	1.5%	100.0%	100.0%	0.0%
Service-producing	21.0%	1.9%	73.6%	76.6%	3.0%
Goods-producing	3.0%	0.4%	26.4%	23.4%	−3.0%
Industrial production	−5.1%	−0.3%	18.9%	15.4%	−3.5%
Information and communication technology	23.9%	2.2%	5.1%	5.4%	0.3%
Manufacturing	−8.0%	−0.6%	15.5%	12.2%	−3.2%

Table 3. Top 10 manufacturing industries in Ontario, Canada by value ($CAD) and # of employees (full- and part-time), 2017. Statistics Canada, 36-10-0402-01 and Statistics Canada, 14-10-0202-01.

Ontario (2017)			
Sector	Value (x1,000,000)	Sector	Employees
Transportation equipment manufacturing	$17,386	Transportation equipment manufacturing	127,965
Food manufacturing	$10,540	Food manufacturing	78,587
Chemical manufacturing	$8368	Fabricated metal product manufacturing	65,228
Machinery manufacturing	$8174	Machinery manufacturing	61,101
Fabricated metal product manufacturing	$6466	Plastics and rubber products manufacturing	49,736
Plastics and rubber products manufacturing	$4845	Chemical manufacturing	46,738
Primary metal manufacturing	$4395	Computer and electronic product manufacturing	29,763
Petroleum and coal product manufacturing	$3971	Furniture and related product manufacturing	29,365
Computer and electronic product manufacturing	$3235	Primary metal manufacturing	26,490
Beverage and tobacco product manufacturing	$2955	Miscellaneous manufacturing	24,493

sectors, each represented over 35% of the province's ICT industry value in 2017. The latter sector represented over 45% of the province's ICT workforce. However, in the same span between 2007 and 2017, employment and output from the manufacturing of computer and electronic products fell by almost one-third.

This is not surprising given the rising weakness in Ontario's manufacturing industry. Between 2007 and 2017, the industry's value fell by 8% and the manufacturing workforce dropped 16%. Table 3 lists the ten manufacturing sectors in Ontario with the largest output and employment in 2017. The 10 largest manufacturing sectors were evenly divided between the number of durable (5) and non-durable manufacturing (5), though the former's output is cumulatively larger than the latter's, driven by manufacturing in transportation equipment.

In the case of Massachusetts's ICT industry, the script is flipped from Ontario's. The perception of the state as a computer manufacturing hub is supported in the data, as computer and electronic product manufacturing produced the bulk of manufacturing GDP and employees in the state in 2017 (see Tables 4 and 5) (Bureau of Economic Analysis, 'Regional data: GDP … '; Bureau of Economic Analysis, 'Regional data: Full-Time

Table 4. Growth rate of industries in Massachusetts from 2007 to 2017. U.S. Bureau of Economic Analysis.

Massachusetts industry growth (2007–2017)					
Industry	Growth	Average growth rate	% of industry share (2007)	% of industry share (2017)	Change in industry share (2007–2017)
All industries	20.3%	1.9%	100.0%	100.0%	0.0%
Service-producing	22.8%	2.1%	83.7%	85.4%	1.7%
Goods-producing	7.8%	0.8%	16.3%	14.6%	−1.7%
Industrial production	11.8%	1.2%	12.6%	11.7%	−0.9%
Information and communication technology	82.5%	6.3%	5.6%	8.4%	2.9%
Manufacturing	10.6%	1.1%	10.9%	10.1%	−0.9%

Table 5. Top 10 manufacturing industries in Massachusetts, USA by value ($USD) and number of employees (full- and part-time), 2017. US Bureau of Economic Analysis.

	Massachusetts (2017)		
Sector	Value (x1,000,000)	Sector	Employees
Computer and electronic product manufacturing	$15,871	Computer and electronic product manufacturing	54,792
Chemical manufacturing	$9616	Fabricated metal product manufacturing	32,962
Miscellaneous manufacturing	$5519	Food manufacturing	27,122
Fabricated metal product manufacturing	$5157	Miscellaneous manufacturing	22,657
Food and beverage and tobacco product manufacturing	$2915	Chemical manufacturing	17,641
Machinery manufacturing	$2123	Machinery manufacturing	17,589
Other transportation equipment manufacturing	$1947	Plastics and rubber products manufacturing	12,814
Electrical equipment, appliance and component manufacturing	$1559	Printing and related support activities	11,662
Plastics and rubber products manufacturing	$1485	Electrical equipment, appliance and component manufacturing	9099
Printing and related support activities	$1230	Paper manufacturing	7882

… '). Between 2007 and 2017, the growth of the state's manufacturing industry has lagged other industries in Massachusetts, with manufacturing as a share of the state's GDP reduced by 0.9% in the time span. However, the output from the computer and electronic manufacturing sector grew by 42% in the same period, increasing its share in the state's GDP by 0.5%.

Massachusetts's ICT industry overall saw positive gains between 2007 and 2017, having grown over 80% in the 10-year time span (averaging 6% per year) and increased its share of the state's GDP by 3%. While the state's manufacturing industry is weakening, the strength of its computer and electronic sector means the ICT industry does not rely as heavily on its technology-related services industry to drive growth as Ontario does. Interestingly, almost a quarter of jobs in Massachusetts manufacturing computer and electronics were lost between 2007 and 2017. This coincides with, and exceeds, the 15% reduction in the state's manufacturing jobs in the same period. While it may seem contradictory to witness growth in manufacturing output but a drop in employment, this may be a signal of strong innovative activity where more economic value is created with less human capital with more efficient processes.

Massachusetts – smart people, doing smart things

At the core of Massachusetts's networks of innovation in technology are four nodes of innovative activities: original equipment manufacturers (OEMs), small and medium enterprises (SMEs), startups and universities play a major role in the state's most important industry clusters. Massachusetts has proven to be an interesting case as manufacturing is integral to several of its most important industry clusters, yet the state is a high wage, high costs state that must compete globally (Reynolds & Uygun, 2018). The state's operational tendencies are perhaps due to its higher education capacities that are among the strongest in the nation (the state produces the highest number of STEM graduates in the United States on a per capita basis), and Massachusetts is home to some of the world's most innovative universities: Massachusetts Institute of

Technology (MIT), Harvard University, Tufts University, Boston University and University of Massachusetts (Massachusetts Technology Collaborative, 2017; Ewalt, 2018). These institutions move beyond the traditional activities of supplying talent and research, as they offer a diverse set of entrepreneurial programs and expertise that enhance commercialization, technology formation, collaborations and financial services. The Conference of Board of Canada's Report on MIT's innovation, for example, highlighted four areas contributing to success: intrapreneurship and entrepreneurship, courses and educational experiences, connections to the research engine and collocated spaces (Dimick, 2014). These four areas highlight how the institution is moving beyond the traditional notions of education and talent supply towards a model that promotes collaboration, reduces risks, offers resources and fosters a regional environment that is conducive to entrepreneurial ventures.

Beyond the educational components of the state's higher education, Massachusetts is also home to intermediaries aimed at supporting entrepreneurship, like the Massachusetts Technology Collaborative. One branch of this publicly funded economic development agency is dedicated exclusively to fostering industry competitiveness through grants, business development programs, networking events, workforce talent education and retention programs. For example, the MassTech Intern Partnership is an initiative that creates a regional talent pipeline between local universities and technology start-ups by offering stipends to small companies that hire local summer interns. This helps promote greater retention rates in the state as well as provide greater work-related experiences for students (Massachusetts Technology Collaborative, 2017).

In Massachusetts, the public sector is present but treads lightly, encouraging the operations of intermediaries such as higher education and research activities. Universities are the unequivocal leader in the state's innovation networks. With a skilled workforce produced and leveraged by its universities, Massachusetts models the potential for endogenous production and retention of innovative practices.

Ontario

Representing 6% of provincial GDP in 2018, Ontario's technology sector is the largest in Canada and valued at $41 billion (CAD) (Statistics Canada, 'Table 36-01-0402-01 … ') (see Appendix B). A multi-nodal effort between government, universities and the private sector, the tech sector is continuing to grow as a component of Ontario's economy. Ontario's regional innovation networks are developed through coordinated efforts between public, private and academic actors to identify regional market trends and create programs intended to establish an environment open to risk and entrepreneurial activities (Galvin, 2019).

Clustered in the cities of Waterloo, Toronto and Ottawa, each area has at least one university and supporting partnerships that connect the public, private and academic sectors. The University of Waterloo has served as a hub for spin-off activities, start-ups and in-migration firms (Bramwell, Nelles & Wolfe, 2008). The university performs its traditional roles of talent supply and research capabilities well, being a staple reason for why firms tend to stay within or migrate to the region. More broadly, universities act as intermediaries, in which they are part of the network of innovative activities. In the Industry 4.0 context, Southwestern Ontario, in particular, is well-positioned to lead advanced manufacturing and Industry 4.0 in Canada and globally, given it is home to a number of industries – including automotive manufacturing – as well as leading academic institutions, start-ups and established companies developing digital tools and solutions (PWC, 2019).

Taking a closer look at one of the province's most innovative regions, Kitchener-Waterloo, the university system offers a conducive environment for tech innovation. With the largest co-op program in Canada, the University of Waterloo actively pursues relations with industry in creating localized knowledge spillovers that are invaluable for regional firms. The university's academic and industry research is well-acknowledged, with the most IT Canada Research Chairs (9) in 2018/19 and receiving a quarter ($53 million) of its research budget from industry (Canadian Association of University Business Officers, 2019). Entrepreneurs are aided by its unique intellectual property (IP) system that allows IP ownership to rest with the creator rather than the institution.

Beyond the higher educational institutions, Waterloo's Communitech, the region's industry association, works to embody entrepreneurship within the region. Communitech not only represents industry interests in policy, but most importantly, provides open collaborative platforms between innovators and entrepreneurs, offering networking and mentoring services (Communitech, 2019). This supportive and collaborative network acts as a strong incentive for firms to stay in the region, even when costs are cheaper elsewhere (Bramwell, Hepburn, N. & Wolfe, 2019). Industry associations and other intermediaries seek to stimulate regional business know-how and entrepreneurial curiosity rather than focusing on creating client interactions. This allows firms to know how to compete globally while simultaneously benefiting from localized knowledge clusters and resources that enhance regional competitiveness (Bramwell, Nelles & Wolfe, 2008).

Shifting the focus to two other tech regions, Toronto's MaRs Discovery District and Invest Ottawa are collaborative platforms that distribute social and material attributes toward the ecosystem, promoting cross-sectoral collaboration. These institutions aim to leverage existing relationships with regional governments, academia and industry actors to stimulate a conducive environment for start-up activities and the creation of regional advantages. In Toronto, the presence of innovative higher education institutions, such as the University of Toronto (which ranked among the world's most innovative universities by Reuters), have aided in supplying necessary talent (Ewalt, 2018). Ontario's fledgling tech sector evidences a multi-nodal approach that integrates the efforts of contributing partners to achieve incremental advances in a massive industry. The opportunity for Industry 4.0 is in the expansion of traditional manufacturing processes into distributed network-based. While it does not rival Silicon Valley's scope or output, nor can it challenge Massachusetts's human capital, Ontario demonstrates the potential for incremental progress by integrating the public, private and university sectors to advance the potential for Industry 4.0 by reducing uncertainty and providing access to subject matter expertise in digital integration.

Discussion

The emergence of information-based production-related networks, such as Industry 4.0, is characterized by the confluence of demand (jobs/industry), supply (labour and expertise) in a more traditional model, and network intermediaries, both physically and virtually contributing to networks of innovation. At the level of the regional structure, and with regard to policy, these cases suggest there is no 'single' silver bullet, guaranteed incentives of innovation, or magic regional policy strategy. Each of the regions and embedded cities have different resources, structures and actors that encourage and foster their networks of innovation. The participation of the public sector – federal, state/provincial and local governments, as well as universities and intermediaries, in different formulations can help to explain the emergence and sustenance of different place-based networks with varying results. Thus, we can conclude that while regional structures are present, and an important factor, it is difficult to tease out in the present cases the 'specific' impact top-down regional policy has on Industry 4.0 and the ability to adapt to new and upcoming opportunities.

There are four takeaways to summarize the lessons of the case studies:

Critical mass of one or more industrial clusters matters: Whether it is natural resources or skilled people (e.g. Massachusetts), network actors can passively utilize existing conditions or move outside of traditional organization boundaries in network/platform-based collaborations characteristic of Industry 4.0. Even Ontario's tech sector was developed around existing universities and cities with critical masses of competently skilled people. It has long been recognized that industrial clusters can be associated with innovative activities, this applies virtually (in this case purely virtual/digital endeavors) as well as those associated with physical, geographic ones. This is, of course, should be expected in a platform-based strategy like Industry 4.0. (Florida, Adler & Mellander, 2017; Frenken, Cefis & Stam, 2015).

Context matters: For policymakers, this means that the context, or regional political/ social/economic environment, is a critical consideration. For example, a conservative-oriented state will likely have little desire, or opportunity for intervention, 'absent disruptive factors'. Unless circumstance provides a window of opportunity, e.g. major crisis, the general approach would appear for the public sector 'to get out of the way' or 'create space' for connections between actors. In this example, the government, as an actor plays a passive role. However, even under this minimalist government philosophy, there are approaches that can enhance the likelihood for innovative economic network outcomes. This could be direct (e.g. contracts and tax incentive programs) or indirectly (e.g. university funding and other network intermediaries) to stimulate innovation.

Collaborative synergies: If starting from nothing or trying to get something started, policymakers should leverage resources and expertise, that must be paired with consistent efforts. Ontario's initial success may be as much about innovative ideas, as it is about government, industry and universities, as well as other actors, working in concert and openly collaborating to move the tech agenda ahead. Too often organizations are rooted in established practices, which frequently are a consequence of existing and perceived limitations. Collaborative policy formation, then, seeks to draw on the varied expertise and resources of network actors to craft solutions that can enhance industrial and production efficacy.

Network Intermediaries: Intermediaries in Massachusetts focus on creating spaces for interactivity. This suggests the importance of proximity of people and ideas in generating innovation and firm development via the operation of network intermediaries. In Toronto, intermediaries such as Toronto Region Board of Trade and the MaRS Discovery

District, serve to aggregate knowledge and increase diffusion on knowledge and best practices. A key role here serves to both enhance actor connectivity as well as serve as a signaling mechanism to the public sector of (1) interest from industry and (2) potential opportunities that might not be currently addressed by policy. We see here that there is a difference in the way that network intermediaries operate, which is in part a function of the divergent strategies of regions and existing economic patterns.

Insights from an application of network and expertise based innovation adoption approaches can help guide (1) design of policy that addresses the complexities of the changing forms of industry, research and development, as well as (2) actively support the growth of competitive industrial knowledge bases, needed by the demands of information economies. From an outcome standpoint, the adoption of collaborative digital technologies means that industrial innovation is not 'necessarily' linked to a physical place and time, but can also be accomplished asynchronously and virtually. While this has become fairly standard in knowledge work, production and manufacturing processes are also subject to these changes, evident in the advent of Industry 4.0. It does, however, suggest that a region needs to actively develop or foster a business case or rationale for a given enterprise or cluster of industries, to locate, or expand in a locale.

Given regional variation in expertise, industrial capacity and research sectors, an interesting question is not 'what are the static policy options that can foster such networks of innovation and, consequently, economic growth and jobs within the Industry 4.0 context', but what policy-related solutions can be generated that collaboratively build on the concept of regional innovation activities? For policymakers, it's critical to (1) be able to leverage data collection and tools of digital integration, from traditional geographic regional sources as well as from industrial sources, as well as (2) develop a robust understanding of the flow of knowledge and intangible assets within regional innovation networks. Governments want to know what they can do (or not do) to help this process along, and how collaborative partnerships and expertise networks are nurtured. Consider, for instance the software sector explored in Massachusetts and Ontario. These regional nodes of innovation are interesting cases of sector-specific trends that represent variation in the type and networks of innovative processes and technologies, and interesting applications for Industry 4.0 as an extended concept.

The public sector – in this case, provincial and state governments – have played an important role in direct investing in and incentivizing innovation. Ontario has actively invested in its technology sector, trying to foster a growing industry with grants for accelerators and incubators that encourage connections between industry, universities and government. By contrast, the role of the public sector is somewhat less direct in Massachusetts, where the government, acting in concert with other stakeholders such as the university sector (both public and private) and other intermediaries, plays a critical role not only in research and training but also in facilitating networks of innovation.

Conclusion

Regional characteristics and advantages can make tangible impacts on the emergence and efficacy of industry innovation in comparatively short periods of time by: (1) assessing the capacity of existing actors and geographic advantages; (2) engaging with broadly representative sets of stakeholders, both internally and externally, to come up with desirable outcomes and economic scenarios; (3) encouraging and supporting cross-sector collaborations (horizontally) and intra- and inter-regional networks of knowledge and (4) devising policy that indirectly, for example through incentives, rather than directly drives innovation. While we see what has been, to date, a shift from the production of 'things' in a traditional manufacturing sense to systems of production – where the services and supporting software have taken a predominant role. An additional consequence has been a shift from labour-intensive manufacturing processes to one where digital technologies amplify a more highly skilled technical workforce.

The digital basis of Industry 4.0 and of modern networks of innovation and production allow the participation of a broader array of actors – such as network intermediaries who provide additional nodes of connectivity and channels for flow of knowledge. The recent global COVID-19 pandemic has proven to be an interesting natural experiment in disruption. Patchy shortages of goods have appeared as the production of several commodities has become centralized in lower cost countries. Interestingly enough, the shift of knowledge workers from centralized offices to remote work has gone relatively smoothly, suggesting that, in many countries, the ICT infrastructure is sufficient. What has been more problematic is the situation that many types of manufacturing still require in-person presence to operate, but at the cost of relatively higher rates of infection in the workforce. However, an Industry 4.0 orientation can be expected to allow more rapid shift in product mixes and resource allocation.

Before the COVID-19 pandemic, policy response in both the United States and Canada have been primarily at a federal level, which has been to allocate fairly large amounts of money to directly support the continued operation of manufacturing and industry. A second response has been to allocate fairly significant sums to support various types of technical and scientific research to address the pandemic. However, the lack of clear federal innovation policy in the US made apparent the need for state-level response to facilitate industrial sustainability. Finally, in terms of local network innovation, the exact impact is not yet discernable, but the short-term impact has been to significantly shift the focus of knowledge flow into pandemic-related inquiry. The longer-term impacts on innovation cannot yet be estimated, but can be expected to result in a shift in focus to biomedical and health-related innovation, as well as a greater attention on the insights resulting from the reliance on digital transformation and flexible, adaptable Industry 4.0 approaches to production.

Our analysis suggests a number of different dimensions that might yield interesting findings. These include, for instance, a deeper larger case examination of the relationship between top-down policy approaches and the presence (or even formation) of regional innovation ecosystems as compared to bottom-up market and stakeholder-driven approaches. Here there is possible a chicken and egg situation, – do policy initiatives form in the presence of nodes of industrial and research innovation, or can top-down policy initiatives stimulate innovative enterprise, absence of any number of associated conditions? Furthermore, more additional empirical research can be done to examine the specific mechanisms of the operation of network intermediaries in engagement with industry actors, governments and academia to generate environments conducive to innovation and economic development.

Note

1. The sectors in the ICT industry was classified according to the classification from Statistics Canada. It combines the North American Industry Classification System (NAICS) code 334 excluding 3345, 4173, 5112, 517, 518, 5415, 8112.

Disclosure statement

No potential conflict of interest was reported by the author(s).

References

Barrutia, J., C. Echebarria, V. Apaolaza-Ibáñez, and P. Hartmann. 2014. "Informal and Formal Sources of Knowledge as Drivers of Regional Innovation: Digging a Little Further Into Complexity." *Environment and Planning A* 46 (2): 414–432. doi:10.1068/a462

Bramwell, A., N. Hepburn, and D. Wolfe. 2019. "Growing Entrepreneurial Ecosystems: Public Intermediaries, Policy Learning, and Regional Innovation." *Journal of Entrepreneurship and Public Policy* 8 (2): 272–292. doi:10.1108/JEPP-04-2019-0034.

Bramwell, A., J. Nelles, and D. Wolfe. 2008. "Knowledge, Innovation and Institutions: Global and Local Dimensions of the ICT Cluster in Waterloo, Canada." *Regional Studies* 42 (1): 101–116. doi:10.1080/00343400701543231

Bureau of Economic Analysis (BEA). n.d. Regional Data: GDP in Current Dollars (SAGDP2) [Dataset]. Accessed August 16, 2020. Retrieved from https://apps.bea.gov/itable/iTable.cfm?ReqID=70&step=1

Bureau of Economic Analysis (BEA). n.d. Regional Data: Full-Time and Part-Time Employment by Industry (SAEMP25) [Dataset]. Accessed August 16, 2020. Retrieved from https://apps.bea.gov/itable/iTable.cfm?ReqID=70&step=1

Canadian Association of University Business Officers. (2019). Financial Information of Universities and Colleges: 2017-18. Retrieved from https://www.caubo.ca/knowledge-centre/surveysreports/fiuc-reports/#squelch-taas-accordion-shortcode-content-6

Chang, Y., and M. Hughes. 2012. "Drivers of Innovation Ambidexterity in Small- to Medium-Sized Firms." *European Management Journal* 30 (1): 1–17. doi:10.1016/j.emj.2011.08.003

Clark, J. 2014. "Sitting 'Scientific Spaces' in the US: The Push and Pull of Regional Development Strategies and National Innovation Policies." *Environment and Planning C: Government and Policy* 32 (5): 880–895. doi:10.1068/c1271r

Clayton, P., M. Feldman, and N. Lowe. 2018. "Behind the Scenes: Intermediary Organizations That Facilitate Science Commercialization Through Entrepreneurship." *Academy of Management Perspectives* 32 (1): 104–124. doi:10.5465/amp.2016.0133

Coenen, L., B. Asheim, M. Bugge, and S. Herstad. 2017. "Advancing Regional Innovation Systems: What Does Evolutionary Economic Geography Bring to the Policy Table?" *Environment and Planning C: Politics and Space* 35 (4): 600–620. doi:10.1177/0263774X16646583

Communitech. (2019). About Communitech – Communitech. Accessed March 11, 2019. Retrieved from https://www.communitech.ca/who-we-are/

Creutzberg, T. 2011. *Canada's Innovation Underperformance: Whose Policy Problem Is It?* Toronto, ON: Mowat Centre for Policy Innovation.

Davis, J., T. Edgar, J. Porter, J. Bernaden, and M. Sarli. 2012. "Smart Manufacturing, Manufacturing Intelligence and Demand-Dynamic Performance." *Computers & Chemical Engineering* 47: 145–156. doi:10.1016/j.compchemeng.2012.06.037

De Silva, M., J. Howells, and M. Meyer. 2018. "Innovation Intermediaries and Collaboration: Knowledge–Based Practices and Internal Value Creation." *Research Policy* 47 (1): 70–87. doi:10.1016/j.respol.2017.09.011

Dimick, S. 2014. *Driving Creativity and Commercialization: Innovation by Design*. The Conference Board of Canada. Retrieved from https://www.conferenceboard.ca/e-library/abstract.aspx?did=6637

Ewalt, D. 2018, October. *Reuters Top 100: The World's Most Innovative Universities—2018*. Reuters. Retrieved from https://www.reuters.com/article/us-amers-reuters-ranking-innovative-univ/reuters-top-100-the-8worlds-most-innovative-universities-2018-idUSKCN1ML0AZ

Florida, R., P. Adler, and C. Mellander. 2017. "The City as Innovation Machine." *Regional Studies* 51 (1): 86–96. doi:10.1080/00343404.2016.1255324

Frenken, K., E. Cefis, and E. Stam. 2015. "Industrial Dynamics and Clusters: a Survey." *Regional Studies* 49 (1): 10–27. doi:10.1080/00343404.2014.904505

Frølund, L., F. Murray, and M. Riedel. 2017. *Engaging in Regional Innovation Ecosystems: Six Questions to Get Your University Partnerships Right!*. MIT Lab for Innovation Science and Policy Working Paper. Retrieved from MIT Innovation Initiative https://innovation.mit.edu/assets/Engaging-in-Regional-Innovation-Ecosystems.pdf

Gale Publishing. n.d. Associations Unlimited. Retrieved from https://www.gale.com/c/associations-unlimited

Galvin, P. 2019. "Local Government, Multilevel Governance, and Cluster-Based Innovation Policy: Economic Cluster Strategies in Canada's City Regions." *Canadian Public Administration* 62 (1): 122–150. doi:10.1111/capa.12314.

Gaspard, H., P. Baker, and J. Zhu. 2019. "How Regions Become Innovation Hubs." *The Hill Times*, 28. Wednesday, Feb, 13, 2019. https://www.hilltimes.com/2019/02/13/regions-become-innovation-hubs/188212

Hilpert, U. 2019. "Systematics and Opportunities of Diversities of Innovation." *Diversities of Innovation* 220: 323–333. doi:10.4324/9781315641836

Hilpert, U., and H. Lawton Smith. 2013. *Networking Regionalised Innovative Labour Markets*. London and New York: Routledge Ltd.

Innovation, Science and Economic Development Canada (ISED). (2018). Report from Canada's Economic Strategy Tables: The Innovation and Competitiveness Imperative. Retrieved from https://www.ic.gc.ca/eic/site/098.nsf/eng/h_00020.html

Kivimaa, P., W. Boon, S. Hyysalo, and L. Klerkx. 2017. *Towards a Typology of Intermediaries in Transitions: a Systematic Review*. Science Policy Research Unit. Retrieved from https://papers.ssrn.com/sol3/papers.cfm?abstract_id=3034188

Kramer, J., E. Marinelli, S. Iammarino, and J. Diez. 2011. "Intangible Assets as Drivers of Innovation: Empirical Evidence on Multinational Enterprises in German and UK Regional Systems of Innovation." *Technovation* 31 (9): 447–458. doi:10.1016/j.technovation.2011.06.005

Massachusetts Technology Collaborative. (2017). FY 2017 Impact Report. Retrieved from http://masstech.org/sites/mtc/files/documents/MassTech/FY17ImpactReport-Web.pdf

Mittal, S., M. Khan, D. Romero, and T. Wuest. 2019. "Smart Manufacturing: Characteristics, Technologies and Enabling Factors." *Proceedings of the Institution of Mechanical Engineers, Part B: Journal of Engineering Manufacture* 233 (5): 1342–1361. doi:10.1177/0954405417736547

Muro M., and Liu, S. 2017. Tech in metros: The strong are getting stronger. Retrieved from https://www.brookings.edu/blog/the-avenue/2017/03/08/tech-in-metros-the-strong-are-getting-stronger/

Muro, M., J. Parilla, G. Spencer, D. Kogler, and D. Rigby. 2018. *Canada's Advanced Industries: A Path to Prosperity*. Washington, DC: The Brookings Institution and The Martin Prosperity Institute, Rotman School of Management, University of Toronto. Retrieved from https://www.brookings.edu/wp-content/uploads/2018/06/Canadas-Advanced-Industries_18-06-05_FINAL2.pdf

National Center for Education Statistics, U.S. Department of Education. 2019. Retrieved from https://nces.ed.gov/fastfacts/display.asp?id=84

Subcommittee on Advanced Manufacturing Committee on Technology of the National Science & Technology Council. 2018. Strategy for American Leadership in Advanced Manufacturing. USA. Retrieved from https://www.manufacturing.gov/sites/default/files/2021-06/Advanced-Manufacturing-Strategic-Plan-2018.pdf

O'Gorman, B., and W. Donnelly. 2019. "Contextualisation of Innovation: The Absorptive Capacity of Society and the Innovation Process." In *Diversities of Innovation*, edited by U. Hilpert, 296–320. Routledge.

Ontario Ministry of Training, Colleges, and Universities. n.d. Retrieved from https://www.ontario.ca/page/ontario-colleges

President's Council of Advisors on Science and Technology. 2014. Report to the President: Accelerating U.S. Advanced Manufacturing. Retrieved from https://obamawhitehouse.archives.gov/sites/default/files/microsites/ostp/PCAST/amp20_report_final.pdf

Pugh, R., W. Lamine, S. Jack, and E. Hamilton. 2018. "The Entrepreneurial University and the Region: What Role for Entrepreneurship Departments?" *European Planning Studies* 26 (9): 1835–1855. doi:10.1080/09654313.2018.1447551

PWC. 2019. Thinking Like a Digital Champion: How Canadian Companies Can Embrace Industry 4.0. Retrieved from https://www.pwc.com/ca/en/industries/building-the-digital-enterprise-key-findings-global/embrace-industry-4-0.html

Ranasinghe, A. 2017. "Innovation, Firm Size and the Canada-U.S. Productivity gap." *Journal of Economic Dynamics and Control* 85: 46–58. doi:10.1016/j.jedc.2017.09.004

Reynolds, E., and Y. Uygun. 2018. "Strengthening Advanced Manufacturing Innovation Ecosystems: The Case of Massachusetts." *Technological Forecasting and Social Change* 136: 178–191. doi:10.1016/j.techfore.2017.06.003

Schwab, K. 2016. *The Fourth Industrial Revolution*. Geneva: The World Economic Forum.

Statistics Canada n.d. Table 14-10-0202-01 Employment by Industry, Annual [CANSIM data]. Accessed on August 16, 2020. Retrieved from https://www150.statcan.gc.ca/t1/tbl1/en/cv.action?pid=1410020201

Statistics Canada n.d. Table 17-10-0005-01 Population Estimates on July 1st, by Age and Sex [CANSIM data]. Accessed on August 16, 2020. Retrieved from https://www150.statcan.gc.ca/t1/tbl1/en/tv.action?pid=1710000501

Statistics Canada. n.d. Table 36-10-0402-01 Gross Domestic Product (GDP) at Basic Prices, by Industry, Provinces and Territories (x 1,000,000) [CANSIM data]. Accessed on August 16, 2020. Retrieved from https://www150.statcan.gc.ca/t1/tbl1/en/tv.action?pid=3610040201

Statistics Canada. n.d. Table 37-10-0130-01 Educational Attainment of the Population Aged 25 to 64, by Age Group and Sex, Organisation for Economic Co-operation and Development (OECD), Canada, Provinces and Territories [CANSIM data]. Accessed August 16, 2020. Retrieved from https://www150.statcan.gc.ca/t1/tbl1/en/cv.action?pid=3710013001

Tiffin, S., and M. Kunc. 2011. "Measuring the Roles Universities Play in Regional Innovation Systems: a Comparative Study Between Chilean and Canadian Natural Resource-Based Regions." *Science and Public Policy* 38 (1): 55–66. doi:10.3152/016502611X12849792159317

Wells, S., and G. Hungerford. 2011. "High-growth Entrepreneurship: The Key to Canada's Future Economic Success." *Policy Options* 31 (8), Retrieved from http://policyoptions.irpp.org/magazines/innovation-nation/high-growth-entrepreneurship-the-key-to-canadas-future-economic-success/

World Economic Forum & Kearney, A.T. 2018. *Readiness for the Future of Production Report 2018*. World Economic Forum. Retrieved from http://www3.weforum.org/docs/FOP_Readiness_Report_2018.pdf

United States Census Bureau. n.d. Educational Attainment: Massachusetts [data from TableID S1501]. Retrieved from https://data.census.gov/cedsci/all?q=bachelors%20degree%20massachusetts

United States Census Bureau. n.d. ACS Demographic And Housing Estimates: United States [data from TableID DP05]. Retrieved from https://data.census.gov/cedsci/all?q=dp05

Industry 4.0 as a 'sudden change': the relevance of long waves of economic development for the regional level

Walter Scherrer

ABSTRACT
While the rapid proliferation of the Industry 4.0 concept suggests that it brings about major economic change, the concept lacks a socio-economic foundation of change. In this context, the paper raises three questions: First, how *sudden* does the change brought about by Industry 4.0 occur, and, second, does it make a difference if it is conceived as *sudden* or not? Using a model of long waves of economic development and the concept of general purpose-technologies it is argued that much of the change brought about by Industry 4.0 is not of a sudden nature but largely represents the roll-out of the prevailing techno-economic paradigm, and that digitalization and artificial intelligence have the potential to trigger of a next long wave. The third question asks for possible implications of the long-wave perspective for adapting to an upcoming paradigm at the regional level. A major conclusion is that policy measures ought to reach beyond traditional innovation and technology-related policies and be implemented in the appropriate phase of a long wave.

1. Introduction

It took only 5 years that the number of hits in Google Scholar that contains the expression 'Industry 4.0' (or its German version 'Industrie 4.0') in the title of scholarly publications increased from zero to more than 1000 per year. Given the quick proliferation of this expression, it is fair to say that the concept of Industry 4.0 has rather 'suddenly' received a prominent place in academia (industry, consultants and politics are fascinated by the concept anyway). Proponents of the concept argue that mankind is at the beginning of a fourth industrial revolution which will bring about fundamental changes to the way how business and production will be done and how our lives will be organized. While it has been made clear that it would take Industry 4.0 some time to revolutionize the economy as a whole, the change it brings about – at least in manufacturing – would be unavoidable and disruptive in nature.

Within this context, this essay tries to answer the following questions:

- First, how 'sudden' has the concept Industry 4.0 emerged, how 'sudden' is the actual and possible change experienced that it is likely to bring about, and could the perception of 'suddenness' of change vary across regions?
- Second, does it make a difference for economic policy if Industry 4.0 is conceived as bringing about a 'sudden' change or not?
- Third, what are regional and policy implications of interpreting Industry 4.0 in a long wave model?

These questions will be discussed in a model of long waves of economic development that does not consider technological change as exogenous influence but analyses it in a wider socio-economic context. This distinguishes our approach from the literature on Industry 4.0, that originates from a policy paper about sustaining the competitiveness of the German manufacturing sector with a strong focus on technology and its ramifications with the business sector (Kagermann, Wahlster, and Helbig 2013). Propagators of Industry 4.0 tell a story of successive industrial revolutions[1] in which each revolution stands for itself, and it is claimed that now the time has come for another such revolution. As there is not much of a socio-economic foundation of 'industrial revolutions' provided in this story it represents what we call the 'stand-alone' perspective of the Industry 4.0 concept.

In contrast to this perspective, we interpret the change that Industry 4.0 brings about in the Perez (2002) model of long waves of economic development. Thereby it is possible to endogenize important aspects of the technological change that is fueled by Industry 4.0, and to analyse the timing of the concept's emergence, changes that it is likely to bring about, and consequently the 'suddenness' of change and the perception of the suddenness of that change. It is argued that while there actually 'is' some sudden change and some changes may be 'perceived' as sudden, most of the change that is involved with Industry 4.0 can be identified as 'not-so-sudden' in nature but as regular phenomena of long-term economic dynamics that is captured by the long wave-concept.

In Section 2 criteria are proposed for conceiving a change as a 'sudden' one, and it is analysed if the stand alone-perspective of the Industry 4.0 concept meets these criteria. Section 3 interprets the Industry 4.0 concept in the socio-economic context of the Perez (2002) model, asks if in this waves-setting the criteria for 'sudden change' apply to Industry 4.0, and asks how regional differences in the suddenness of change and its perception emerge. Section 4 addresses implications for economic policy that derive from interpreting Industry 4.0 in a long wave-perspective in three policy areas (distribution, stabilization and regional policy). Finally, the paper concludes that the Industry 4.0 concept has not emerged suddenly and that the 'perception' of suddenness varies across regions if a long-wave perspective is applied.

2. The 'stand-alone' perspective: Industry 4.0 as a 'sudden' change

This section develops criteria that have to be met for a change constituting a 'sudden' change and analyses if the Industry 4.0 concept meets these criteria. 'Sudden' has several meanings (Merriam-Webster, 2018-10-04): something is (a) happening or coming unexpectedly, changing angle or character all at once, (b) made or brought about in a short time, or (c) marked by or manifesting abruptness or haste. Synonyms for sudden are abrupt, unanticipated, unexpected, unforeseen, unlooked-for. In the context of Industry 4.0 meaning (a) could be relevant because the concept claims changing the character of the economy fundamentally within a rather short period, meaning (b) could be relevant because of the quick proliferation of the concept, and meaning (c) could also be relevant in the sense of abruptness of change. In the context of long-term economic development and fundamental technological and economic change (which is claimed to be brought about by Industry 4.0 when considered as a stand-alone concept), we argue that the following four criteria have to be met for a change constituting a 'sudden' change.

'Relevance of change': First, the impact of change has to be sufficiently strong to impose an impact on doing business and organizing life in virtually all sectors of the economy, all regions, and all realms of society. Otherwise change is not important enough to be considered in the context of long-term structural change and economic development. Proponents of Industry 4.0 as a stand-alone concept claim that it will bring about encompassing change in economy and society:

> The continuing convergence of the real and the virtual worlds will be the main driver of innovation and change in *all sectors of our economy*. The exponentially growing amount of data and the convergence of different affordable technologies that came along with the definite establishment of Information and Communication Technology are transforming *all areas* of the economy. (Kagermann 2015, emphasis by the author)

Yet, the form and intensity of change will differ across regions because of regionally different circumstances and because of diversities of innovation across sectors that impact the adaptability of the regional economy (O'Gorman and Donnelly 2020).

'Disruptiveness of change': Second, the more disruptive a change is, the more likely it will be considered a sudden one. Disruptiveness means that a break is required with institutions, economic structures, habits and other fundamental elements in society which have been well established before over a long period. A major claim of proponents of Industry 4.0 as a stand-alone concept is that it has the potential to bring about not only localized change in some specific markets and industries ('first order disruption'; see Schuelke-Leech 2018) but that it affects many industries and changes social norms and institutions substantially ('second order disruption'). Business models, economic structures, institutions and other fundamental elements in society which have been well established for a long time will (have to) undergo major changes and will be felt in all regions (although differently, again). This argument is extensively elaborated in the literature and need not be further extended here.

'Unexpectedness of change': Third, a change's nature is the more sudden the less it is part of an agent's mindset. If there is no probability known for the occurrence of change and even more so if the set of possible changes is not known, uncertainty about the future increases and (nearly any) change could become considered as 'sudden' in its nature. In contrast, if the set of possible changes is known and probabilities can be attached to each of the possible changes then it is less likely that change will be considered as unexpected. Nevertheless, even in such a situation the timing of the actual occurrence of change and the timing of the unfolding of the change requirements may still be unclear ex ante so that change might still come as a 'surprise'.

If Industry 4.0 is considered as a stand-alone concept, then it claims to be a revolution that requires and ultimately brings about change in all spheres of life – also in those in which change comes largely unexpected. Technological properties, possibilities and implications of Industry 4.0 for individuals and the socio-economic framework are still unclear, and many of the concept's implications and possibilities are not part of most individuals' mindsets. While the emergence and proliferation of Industry 4.0 is likely to affect all regions in some way the (un-)expectedness of the concept's appearance will vary across regions. Firms and the population in regions with a large share of developers and users of Industry 4.0 concepts might be less surprised by the occurrence of change that is brought about by Industry 4.0 compared to regions that need to import these concepts and applications. Summing up, the change brought about by Industry 4.0 can be considered largely a 'sudden change' on the dimension of expectedness if it is considered a stand-alone concept.

'Perception of change': Fourth, for a change to be considered a sudden one it is required that it is also 'perceived' as a sudden change. Therefore, the perception of change will largely depend on the communication of information about it, and on the willingness and ability of the population to adapt to change. The announcement or forecast of change which is likely to come about confronts individuals with uncertainty and at least implicitly requires them to change their habits, lifestyles, ways of earning income and doing business. On the one hand, given a high degree of firms' and individuals' adaptability to change this could be considered as a business opportunity or as an opportunity to improving one's wellbeing. But, on the other hand, as change requirements might be substantial and costly, the proper timing of response is not clear in advance, less well adaptable individuals and those who are less willing to adapt might assume (or hope) that the personal impact of such change might materialize only in the distant future. But not responding to change requirements just delays the necessity to adapt; change requirements will pile up, and ultimately change can be perceived as a sudden phenomenon although it had been forecasted long time before. As adaptability and willingness to adapt is likely to vary across regions (Hu and Hassink 2016), the perception of change might differ across regions.

Perception patterns might be reinforced if change is assumed to be unavoidable – proponents of Industry 4.0 argue that it 'will come inevitably, whether we want it or not' (Drath and Horch 2014, 58). Using branded patterns of argumentation under the label Industry 4.0 has facilitated the concept's reception in the general public, in business and in academia. The label has become so convincing and powerful because it suggests that a new era is going to arrive which has had only three predecessors in history (three former 'technological revolutions'), because it reduces complexity, because it merges a broad variety of topics into one single label, and because the change that it brings about is claimed to be 'unavoidable'.

In conclusion, from a 'stand-alone' view Industry 4.0 meets the four criteria of 'suddenness'. Yet, this view is silent about the causation of that revolution: a few descriptive references to Schumpeterian-type innovation and/or superficial references to Kondratieff waves are not sufficient for providing a sound theoretical foundation for telling the story of a 'revolution'. This view considers industrial revolutions merely as phenomena which just happen to happen from time to time: a revolution is going to happen right now because the appropriate technologies are available right now, and because these technologies are more economic than the existing technologies. While this may not be wrong it tells only part of the story: A more thorough socio-economic framework is required for explaining the emergence of such revolutions, why such a revolution is going to occur right now and thereby giving an answer to the question if this is occurring suddenly.

3. A long wave-perspective: the not-so-sudden appearance of (much of) Industry 4.0

After having discussed Industry 4.0 as a stand-alone concept now it will be interpreted within the Perez (2002) model of technology-driven long waves of economic development. The general analysis of the four criteria developed in the preceding section (relevance, disruptiveness, unexpectedness and perception of change) will be extended to a regional perspective in the final sub-section. Long waves are triggered by general purpose technologies (GPT) that have the capacity to profoundly transform the economy and that ultimately establish a new 'techno-economic paradigm' (TEP) (Perez 1983). The dynamics in the model is driven by 'real economy' entrepreneurs and financial investors, and economic development is the result of an interplay of technological, economic, political and social influences. A technology qualifies as a GPT if it has an impact on all aspects of the economy and society, and if it causes innovation and structural change in nearly all sectors of the economy. The change it brings about means more than just using a new technology, but it means establishing a new best-practice model of organizing business and the economy according to the properties of the new GPT. The label 'industrial revolution' suggests that the proponents of Industry 4.0 assume the concept to constitute a GPT.

A TEP runs through several phases of a long wave (Perez 2002): it starts with the upcoming phase and the 'Frenzy' phase (together forming the 'installation phase' of a new TEP), and ends with the synergy phase and the maturity phase (forming together the 'deployment phase' of a new TEP). Installation and deployment phases are connected by a period of transition ('turning point') that is particularly important from a policy perspective. The current fifth long wave of economic development has been triggered by new (=digital) information and communication technologies (ICTs); at the begin of the 2020s it is in its deployment phase. In order to scrutinize the 'suddenness' of change that is likely to be triggered by industry 4.0 the concept will be interpreted in a long wave context based on the four criteria of 'suddenness' of change.

3.1. Relevance and disruptiveness of change

For clarifying if Industry 4.0 meets the 'relevance' criterion for constituting a sudden change also from a long-wave perspective, the results of two recent systematic surveys of the academic literature are referred to. First, Liao et al. (2017) found 10 meaningful noun phrases which are commonly recognized terms that are associated with Industry 4.0. These are Cyber Physical Systems, Smart Factories, Industrial Revolutions, Internet of Things, Production Systems, Industrial Internet, Manufacturing Systems, Smart Manufacturing, Production Processes and Cyber Physical Production Systems. Second, according to a bibliometric search by Trotta and Garengo (2018) research on Industry

4.0 addresses Big Data Analytics, Cloud Services, 3D-Printing, Cyber Security, Autonomous Robots, Internet of Things, Augmented Reality, Simulation, and Horizontal and Vertical Integration.

Therefore, Industry 4.0 comprises a broad range of technologies that do not only have an impact in a single industry or small group of industries in which the technology initially emerged but on the economy as whole. Regions that are strong in Industry 4.0 applications will feel the economic impact more directly than regions with less Industry 4.0 penetration. The latter ones tend to become vulnerable to advances in digitalization that occurs in more developed regions and, therefore, will feel the impact indirectly as their competitiveness tends to deteriorate (Hickie 2020). By implication, the relevance criterion for Industry 4.0 constituting a sudden change in economic development can be regarded as fulfilled.

If Industry 4.0 meets the criterion of being 'disruptive' within a long wave-perspective depends on whether it qualifies as a GPT: does Industry 4.0 have the power to create a new techno-economic paradigm that changes the ways how business and social life is organized? Two caveats need to be considered: first, not all technological and organizational change taking place currently is related to Industry 4.0. Second, many changes that involve elements of Industry 4.0 are not of a disruptive nature but improve efficiency or performance within existing business models. While the former caveat is self-explanatory the latter one needs some explanation that is based both on theoretical considerations and on empirical observation.

Within the Perez' long-wave model of economic development many Industry 4.0 applications represent the deployment phase of the fifth wave that has been coined by the emergence and proliferation of ICT. In the deployment phase the existing ICT-based TEP is rolled out to ever more firms and branches, particularly in the manufacturing sector (Scherrer 2019). This argument is supported by Drath and Horch (2014, 58) – two proponents of the Industry 4.0 concept – who observe that Industry 4.0 is similar to the ICT penetration of 'the consumer world, which was confronted with the internet in the early 1990s'. In the same vein, Ibarra, Ganzarain, and Igarta (2018, 4) posit in a review essay on Industry 4.0 that '(t)he increasing fusion of industrial production and ICT has brought the so-called Industry 4.0 into the manufacturing world' and thus support the idea that (much of) Industry 4.0 is concerned with the roll-out of the ICT-based TEP. Xu, Xu, and Li (2018, 2942) emphasize a different aspect of this roll-out process in the manufacturing sector and declare that 'Industry 4.0 focuses more on the end-to-end digitization and the integration of digital industrial ecosystems by seeking completely integrated solutions.'

Empirical analyses and case studies of the impact of Industry 4.0 applications are also supportive to the view that many applications represent the roll-out of the ICT-based TEP. E.g. Gaddi, Garibaldo, and Garbellini (2020) find in an in-depth analysis of case studies on Industry 4.0 use in Italian manufacturing firms that these applications were only in few cases disruptive in nature. This finding is confirmed by several in-depth case studies on the experience of internationally operating firms that are headquartered in southern Germany and Austria and operate in the manufacturing and service industries.[2] Two large manufacturers of metal working equipment and a leading manufacturer of industry automation systems report that the adoption of Industry 4.0 applications has not changed their business models. A leading manufacturer of specialized equipment in the health sector has not felt pressure to change its business model, too. Most of its pro-ducts are complex and need much explanation, and product support has been mostly provided by independent specialized distributors. As the firm relies on the distributors it only recently has established a web-shop in cooperation with the distributors for selling standardized goods to the final customer. But again, this does not mean disruption but change that occurs within the existing ICT paradigm.

Regarding the service industries, a major provider of specialized business to business logistics and transport services has not experienced fundamental change of the sector's business model, too. Instead, it is claimed that a significant increase in productivity could be easily achieved if business partners were ready to implement rather simple business procedures based on conventional ICT. A supplier of advanced building services who operates in the business to consumer segment has experienced that building tech-nologies and information technology converge. While ICT has changed the composition of equipment and services provided by the firm the basic business model in the sector has not changed. Finally, a leading provider of real estate services reports that it has started digitizing its huge stock of information and that this process is going to take much more time and effort than expected. It is not yet clear if this process would have a disruptive impact on the firm's business model.

The examples demonstrate that the business models of many existing firms have not (yet) undergone a fundamental change – even in firms that are located in economically advanced regions and that operate in a variety of industries and regions. While all firms report that innovation takes place in the markets in which they operate, in no case a dis-ruptive impact on a firm's business model was found. The lack of reported 'disruptive-ness' could mean that in many firms and sectors innovation brought about by Industry 4.0 would not be disruptive, or that it is only first-order disruption (Schuelke-Leech 2018) within the currently dominating TEP based on and formed by ICT. Therefore – although in all cases it was unclear if, how and when the firm's or sector's business model would be affected by Industry 4.0 in the future – the change brought about so far by innovation based on more intensive use of ICT has not been much of a 'sudden' change but has been in accordance with the long-wave pattern of dis-semination of the dominant TEP.

3.2. Unexpectedness and perception of change

Regarding the dimension of 'unexpectedness', embedding the Industry 4.0 concept into a long wave-model casts also doubts that its appearance marks a 'sudden' change. Here the timing and sequencing of the phases of the current ICT-driven long wave is important. When its turning point-phase began in 2001 after the burst of the speculative bubble ('dot-com Bubble') economic policy was confronted with two challenges. First, after the huge losses in the stock markets, investors' confidence needed to be regained in order to stabilize the economy in the short run. Economic policy successfully contained the short-term direct economic impact of the burst of the ICT-driven speculative bubble through expansionary monetary policy (with some delay in Europe). This encouraged investment primarily in financial capital and not in 'real' capital giving way to the build-up of another speculative financial bubble that was only indirectly related to the dissemination of the ICT-TEP. Its burst in 2008 marked the end of the turning-point phase of the fifth long wave (Perez 2009).

The second challenge in the turning point phase was that the institutional setting needed to be adapted to the requirements of the new TEP. The legal foundations of the internet had been established early, but it took much longer to deal with problems like standardization, privacy and data protection reflecting the slow progress in adjusting the regulatory frame to the needs of the not-anymore-so-new TEP. Only recently IoT-standards – which are required for rolling out ICT particularly in the manufacturing sector – have been developed, and the General Data Protection Regulation in the European Union became only effective in 2017. Beyond technology policy in a narrow sense, adaptations of existing regulations and new regulations were required in many other policy fields (e.g. labour relations, competition policy, social security system) and changing skill requirements needed to be addressed. Different rates of economic growth across regions suggest that the benefits from the ICT-based TEP in the deployment phase of the fifth long wave have not materialized equally across states and regions. While different regional industry structures and clusters that offer different opportunities for rolling out the ICT-based TEP might play a role, substantial differences in economic dynamics among regions of a similar level of economic development suggest that in many regions an appropriate institutional framework has not been found (yet).

According to the phasing of the long wave-model, the deployment phase of the fifth long wave began in the late 2000s by rolling out its ICT-based TEP over the whole economy (Scherrer 2020) by refining, further developing and adjusting these technologies and combining them with other technologies. As this is exactly what many applications of Industry 4.0 are about, from a long wave perspective this kind of innovation and investment has not come unexpectedly. Therefore, it does not constitute sudden change but is within the expected pattern of the long wave dynamics.

The not-so-sudden appearance of Industry 4.0 notwithstanding, economic agents still might 'perceive' change as sudden. Like in the early 1990s when the emergence of the internet caused an 'unpredictable world of online shops, auctions, internet banking, online brokerage, and video streaming' in the consumer world (Drath and Horch 2014, 58), Industry 4.0 is producing surprising results, this time primarily originating from the manufacturing sectors. While there are analytical differences between interpreting Industry 4.0 in a long-wave framework versus treating it as a stand-alone concept, there need not be much difference concerning the 'perception' of change.

Under the powerful label 'Industry 4.0' it has been argued that this concept's implications are not clear ex ante but, nevertheless, the change brought about will be unavoidable. Such a situation is perceived by individuals and firms to be extremely difficult to control, and it is also difficult to develop appropriate responses at the policy level. Some agents might try to neglect or avoid change as long as possible because they have invested substantially in physical and human capital that might be devaluated as a consequence of the changes to come about. Others might not be sufficiently flexible or mobile for adjusting to the change, or consider themselves not being capable of doing so, or may face other restrictions to accommodating change. At the time when they are actually confronted with change and concrete requirements to change, they might be overwhelmed and feel surprised. Therefore, an individual person may perceive change as 'sudden' even if it has been discussed publicly long before it had become effective.

How intensive a change (and the need to change) is perceived depends i.a. on the choices that economic agents have. In historical perspective innovation activity has been concentrated in metropolitan areas (Hilpert 1992, 2016), meaning that both firms and the population in these regions tend to be exposed to the change brought about by Industry 4.0 from its early stages on. This could make economic agents in metropolitan areas perceive that change as less sudden than in regions that are affected by Industry 4.0 only indirectly or later. In conclusion, the criterion of 'perception' for considering change related to Industry 4.0 as sudden change is met only to some degree.

Overall, the provocative conclusion from interpreting Industry 4.0 in a long wave-framework is that the surprise is not that Industry 4.0 as a concept has emerged in the early 2010s but that it did not emerge 'earlier' within the fifth long wave. Much of recent innovation activity is still concerned with rolling out the ICT-TEP over more and more sectors of the economy – which is in line with the long wave model. Therefore, from a long wave-perspective the change has not emerged 'suddenly', although at the same time some change can still be 'perceived' as sudden.

3.3. Industry 4.0 – divergent regional perceptions of disruption in the future?

Both public debate and much of scientific research on Industry 4.0 applications emphasize the productivity increase through automation and robotization and its possibly huge negative impact on employment. While this discussion is relevant, of course, and structural change during the dissemination process of a new technology is enormous, this does not necessarily mean that change is disruptive as long as the productivity impact materializes within existing business models. Thus, it has been argued before that – even in economically advanced regions – many Industry 4.0 applications so far have been representations of the deployment phase of the fifth Kondratieff wave. From a long wave-perspective, in order to qualify as a GPT and, hence, becoming the trigger of a long wave, the changes released by a new GPT need to reach beyond productivity impacts within existing business models and need to affect (nearly) all sectors of the economy. The universal applicability of a GPT to many sectors and the possibility to combine it with many other technologies allows for a large variety of uses in economy and society. Change that is brought about by a new GPT will be considered more likely as a sudden one if a new GPT is of a disruptive nature, and if it is difficult to predict its possible uses and the timing when these uses will eventually become effective.

Regarding Industry 4.0, only those elements that cause 'disruptive' changes in virtually 'all' economic sectors have the potential to constituting a GPT and acting as a trigger of the sixth Kondratieff wave. The strongest candidates are two elements on which Industry 4.0 builds upon: digitalization and artificial intelligence (AI). This is for two reasons: First, based on the conversion of analogue data and processes into a machine-readable format ('digitization') digital technologies can be used that result in new activities or in changes to all kinds of existing activities ('digitalization'; OECD 2018). Both digitalization and AI have the potential for becoming truly 'general' purpose technologies because they are likely to be applicable to all economic sectors and activities, and because they can be combined with many other technologies. Second, applications of digitalization and AI have the potential to disrupt existing business models in many sectors of the economy. In particular the emergence of 'platform economies' with marginal cost of production close to zero becomes possible and creates private monopolies (and undermines the role of marginal cost as a fundamental concept of pricing in a market economy). While regions in which the developers and platform firms are headquartered will benefit directly, the long-wave perspective suggests that other regions have the chance to gain indirectly if they succeed to adopt to the new TEP that will emerge. The economic gains are likely to vary across regions and will depend on the regions' economic structures, and – even more – on their capabilities to adjust both economic structures and the institutional setting.

The ICT industry is likely to continue playing a prominent role as a provider of the infrastructure, and the capabilities and the capacities of internet and mobile communication networks will be further extended and improved (think about IoT and 5G mobile communication). As ICTs' existence has become undisputed and an infrastructural backbone they will act as an enabler for the formation and dissemination of the next GPT. Regional differences in the quality of this infrastructure will be a key determinant of regions' potential for economic and social development in the 6th long wave. Empirical evidence of current innovation processes suggests that firms digitize their processes and thus prepare for the next GPT.[3] Therefore, as digitalization and AI are likely to act as drivers of a sixth long wave and to have the potential to affect business models in many industries, regional differences of the possible economic impact notwithstanding, the disruptiveness criterion is fulfilled. The timing of these technologies' emergence is coherent with the long-wave model because the upcoming of a new TEP is expected to happen during the late part of the deployment phase of the still dominant 'old' TEP.

4. Policy implications from a long-wave perspective: regional effects of national policies and processes

Whether Industry 4.0 is seen as a stand-alone concept or if it is interpreted within a long wave-model of economic development leads to different conclusions on how 'sudden' the change occurs that is going to be brought about by its dissemination. This section discusses if the different perspectives also make a difference for adapting to (not-quite-so-) 'sudden changes' emanating from Industry 4.0 by economic policy makers at different levels of government.

From a stand-alone perspective, economic policy for Industry 4.0 is geared towards technological aspects (e.g. supporting Industry 4.0-specific R&D and innovation) and direct implications of technological change on the business sector (e.g. developing an appropriate skill base for developing and implementing the Industry 4.0 concept). While policy initiatives addressing these fields are important, of course, restricting the policy focus to such a narrow perspective entails two shortcomings: First, it tends to bias the perception of what Industry 4.0 is about and what its implications for economic policy are. Second, harvesting the benefits from the proliferation of Industry 4.0 will require actions beyond the scope of these policy fields. The common cause of the two shortcomings of conceiving Industry 4.0 as a stand-alone concept is a lack of a socio-economic explanation of its emergence and of its dissemination across industries and regions. Policy implications of these shortcomings will be illustrated by referring to distribution policy, to macroeconomic policies, and to regional economic policy.

First, the stand-alone perspective of Industry 4.0 does not conceptualize the nexus between the dissemination of a new technology, regulation, and the distribution of income, and it cannot capture, the impact of divergent industrial structures on regions' economic development potential. The long wave perspective, by contrast, suggests that the distribution of income undergoes systematic influences during the phases of a long wave (Scherrer 2016). In the early phase entrepreneurs start experimenting with the upcoming technology thereby developing the properties of the new TEP. In this phase a policy focus on deregulation can be expected in order to facilitate experimenting with the new technology which, in turn, is conducive for developing new products and services within existing markets, and for creating new markets and activities based on the new technology. Adjustments in taxation – in countries with a multi-level government structure all levels of government are possible actors – are also likely to occur in this phase for providing economic incentives for entrepreneurs and investors to push innovation based on the new technology. By implication, an increasing number of entrepreneurs and investors earn huge profits in short time, while the not successful ones and many of those who do not adapt to the new TEP tend to lose ground. Therefore, inequality both between individuals and regions tends to increase during the Frenzy phase, and the increased inequality tends to be accepted as a consequence of the unfolding of the new TEP. Not surprisingly, values like solidarity are not 'fashionable' in this phase, and it is difficult to get topics onto the political agenda that are in a social-democratic tradition (like reducing inequality or re-distributing income). In the deployment phase of the long wave when paradigms mature, inequality constrains purchasing power and restrains demand for goods and services. As reducing inequality would allow to fully capture the potential benefits of the new TEP the issue of income distribution becomes relevant for reaping the full economic benefits from the TEP. Therefore, economic policy and taxation in particular should emphasize the principle of redistribution in this phase (Freeman, 2011). It is important to recognize that the argument for policies that reduce inequality and foster demand does not rest on social policy considerations but is based on economic grounds.

Second, the need for complementary macroeconomic policies that reach beyond mere technology policy and those policies that address the direct impact of new technologies is not conceptualized in the stand-alone perspective of Industry 4.0, too. Embedding the concept into a long wave-model clarifies the importance of this connection. During the turning point phase and in the early deployment phase there is a 'case for a Keynesian policy of demand' (Reati and Toporowski 2009, 164ff), and in order to prevent countries to fall into deflationary recessions 'the IMF and the World Bank should be brought closer to the original Keynesian ideal' (Freeman, 2011, 23). But after the turning point phase of the current long wave macroeconomic policies have not been conducive to exploiting the economic potentials of the TEP as fiscal policies have been largely procyclical after 2008 (Blanchard and Leigh 2013; Fatas and Summers 2018), and monetary policy has incentivized financial and real estate investment relative to investment in real capital.

Third, while distribution and other macroeconomic policies have indirect implications at the regional level, interpreting Industry 4.0 in a long wave perspective allows also for analysing direct economic implications of the emergence and dissemination of a TEP. Both historical evidence of long waves and the analysis of prior technology developments show that most innovation in cutting-edge technologies has always been concentrated in 'islands of innovation' (Hilpert 1992; 2016). Such places exhibit a strong research and development (R&D) base, and about 90% of such islands of innovation in the United States and in Europe known today have a tradition in related, precursor industries (Hilpert 2016). An analysis of European R&D in those technologies that are summarized under the Industry 4.0 label also shows that the comparative advantage is concentrated in few regions (Ciffolilli and Muscio 2018). Innovation-relevant infrastructure tends to be dominated even more than in former long waves by very few firms that are located in such 'islands'.

But the historic evidence of long waves also shows that for being economically successful regions need not necessarily be R&D champions in those technologies that act as a trigger of a wave. This argument can be illustrated with many regions that adjusted their economies in the fourth Kondratieff wave successfully to the then-dominant TEP without having an own manufacturing base (which was the sector where the mass production TEP initially emerged) by applying the principles of mass production to sectors like tourism (Scherrer 2019). In recent times, in the realm of Industry 4.0 there is also the possibility to initiate innovation by 'crossover' knowledge transfer among recombinant knowledge constellations (Cooke 2018). Existing regional clusters can support the implementation of Industry 4.0 applications if they are equipped with the adequate knowledge base and some expertise in the technology fields that are crucial for Industry 4.0 (Götz and Jankowska 2017). In particular, clusters can be supportive for generating incremental innovation at the regional level (Hervas-Oliver et al. 2019). The gap between regions with different preparedness to Industry 4.0 can be narrowed if regional agents have experience with R&D, if they have a basic understanding of the new technologies, but most of all if they have a good understanding of the new TEP's implications and potential for organizing economy and society. This poses challenges for economic policy at all levels of government in policy areas that reach far beyond the traditional scope of technology, innovation and qualification policy.

Therefore, the analysis of Industry 4.0 in a long wave perspective provides a socio-economic foundation to overcome shortcomings of the stand-alone perspective. This is relevant for economic policy as it allows developing strategies for accommodating the changes that Industry 4.0 can bring about by addressing also the broader and more indirect impact of new technologies. As a consequence, interpreting Industry 4.0 in a long wave model can make change appear less 'sudden' both for policy makers and other economic agents.

5. Conclusion: the not-so-sudden emergence of Industry 4.0 is perceived differently across regions

Interpreting Industry 4.0 within a long-wave model suggests that so far many of its applications have been representations of the roll-out process of ICT over the manufacturing sector and other sectors of the economy. By referring to four criteria that capture the 'suddenness' of change it is shown that much of the change brought about by Industry 4.0 is not of a 'sudden' nature but is in line with the long wave-dynamics. Analyses of policy areas that focus on income distribution, on macroeconomic stability and on the regional economy have illustrated that interpreting the Industry 4.0 concept in a long wave model improves the understanding of the concept and of its dissemination.

From a long wave perspective, digitalization and AI (being parts of the Industry 4.0 concept) very likely form the core of the next TEP. Regional differences in economic structures will have an impact on regions' capabilities to adapt their economies to the requirements of the next TEP. Economic agents and policy makers at national and regional levels need to have a good understanding of the upcoming TEP and the economic potential it provides for appropriately adjusting the institutional framework. It is shown that a range of policy measures that reach beyond the narrow scope of innovation and other policies that are directly related to technology can facilitate adapting to change if they are implemented in the appropriate phases of a long wave. Finally, the long wave-perspective also suggests that, while the emergence of disruptive features of Industry 4.0 – digitalization and artificial intelligence – does not come as a surprise. Nevertheless, actual changes and change requirements that come along with Industry 4.0 might still be 'perceived' as 'sudden' changes, and the perception of suddenness of change is likely to vary across regions. Further research on the triggers of regional variation of the dissemination of a new TEP could improve our understanding of such regional variations and of causes for the (non-) synchronization of long waves across regions.

Notes

1. This story is nicely summarized in Xu, Xu, and Li (2018, 2942):

   > The First Industrial Revolution began at the end of the eighteenth century and early nineteenth century, which was represented by the introduction of mechanical manufacturing systems utilizing water and steam power. The Second Industrial Revolution started in the late nineteenth century, symbolized by mass production through the use of electrical energy. The Third Industrial Revolution began in the middle of twentieth century and introduced automation and microelectronic technology into manufacturing … The introduction of cyber-physical systems will be one of the most revolutionary changes in Industry 4.0 which is described as a fourth industrial revolution.

2. The following examples from the manufacturing and service sectors refer to in-depth in-firm studies from South Germany and Austria that were made accessible to the author between 2017 and 2020.
3. See also most of the examples in Section 3.1.

Disclosure statement

No potential conflict of interest was reported by the author(s).

References

Blanchard, O., and D. Leigh. 2013. "Growth Forecast Errors and Fiscal Multipliers." *American Economic Review* 103 (3): 117–120. doi:10.1257/aer.103.3.117

Ciffolilli, A., and A. Muscio. 2018. "Industry 4.0: National and Regional Comparative Advantages in Key Enabling Technologies." *European Planning Studies* 26 (12): 2323–2343. doi:10.1080/09654313.2018.1529145

Cooke, P. 2018. "Generative Growth with 'Thin' Globalization: Cambridge's Crossover Model of Innovation." *European Planning Studies* 26 (9): 1815–1834. doi:10.1080/09654313.2017. 1421908

Drath, R., and A. Horch. 2014. "Industrie 4.0: Hit or Hype?" *IEEE Industrial Electronics Magazine*, June 2015, 56–58.

Fatas, A., and L. Summers. 2018. "The Permanent Effects of Fiscal Consolidations." *Journal of International Economics* 112: 238–250. doi:10.1016/j.jinteco.2017.11.007

Freeman,, Ch. 2011. "Technology, Inequality and Economic Growth." *Innovation and Development* 1: 11–24. doi:10.1080/2157930X.2010.551062

Gaddi, M., F. Garibaldo, and N. Garbellini. 2020. "The Italian Experience in Implementing Industry 4.0. *UCJC Business and Society Review* 17(2): 52–69. doi:10.3232/UBR.2020.V17.N2.03.

Götz, M., and B. Jankowska. 2017. "Clusters and Industry 4.0 – do They fit Together?" *European Planning Studies* 25 (9): 1633–1653. doi:10.1080/09654313.2017.1327037

Hervas-Oliver, J., F. Sempere-Ripoll, S. Estelles-Miguel, and R. Rojas-Alvarado. 2019. "Radical vs Incremental Innovation in Marshallian Industrial Districts in the Valencian Region: What Prevails?" *European Planning Studies*, doi:10.1080/09654313.2019.1638887.

Hickie, D. 2020. "Diversities of Innovation in Advanced Manufacturing – The Aerospace and Automotive Industries." *UCJC Business and Society Review* 17 (1): 50–65. doi:10.3232/UBR. 2020.V17.N1.03

Hilpert, U. 1992. *Archipelago Europe – Islands of Innovation. Synthesis Report. Prospective Dossier No 1: Science, Technology and Social and Economic Cohesion in the Community*. Brussels: Commission of the European Communities.

Hilpert, U. 2016. "The Culture – Technology Nexus: Innovation, Policy and the Successful Metropolis." In *Routledge Handbook of Politics and Technology*, edited by U. Hilpert, 149–161. Oxford: Routledge.

Hu, X., and R. Hassink. 2016. "Explaining Differences in the Adaptability of old Industrial Areas." In *Routledge Handbook of Politics and Technology*, edited by U. Hilpert, 162–172. Oxford: Routledge.

Ibarra, D., J. Ganzarain, and J. Igarta. 2018. "Business Model Innovation Through Industry 4.0: a Review." *Procedia Manufacturing* 22: 4–10. doi:10.1016/j.promfg.2018.03.002

Kagermann, H. 2015. "Change Through Digitization – Value Creation in the Age of Industry 4.0." In *Management of Permanent Change*, edited by H. Albach, H. Meffert, A. Pinkwart, and R. Reichwald, 23–45. Wiesbaden: Springer Gabler.

Kagermann, H., W. Wahlster, and J. Helbig. 2013. *Deutschlands Zukunft als Produktionsstandort Sichern. Umsetzungsempfehlungen für das Zukunftsprojekt Industrie 4.0. Abschlussbericht des Arbeitskreises Industrie 4.0*. Frankfurt/Main: Acatech – Deutsche Akademie der Technikwissenschaften.

Liao, Y., F. Deschamps, E. de Freitas Rocha Loures, and L. Pierin Ramos. 2017. "Past, Present and Future of Industry 4.0 – a Systematic Literature Review and Research Agenda Proposal." *International Journal of Production Research* 55 (12): 3609–3629. doi:10.1080/00207543.2017. 1308576

Merriam-Webster online dictionary. https://www.merriam-webster.com/dictionary/sudden 2018-10-04)

OECD. 2018. *Going Digital in a Multilateral World*. Paris: OECD.

O'Gorman, B., and W. Donnelly. 2020. "Contextualisation of Innovation. The Absorptive Capacity of Society and the Innovation Process." In *Routledge Handbook of Politics and Technology*, edited by U. Hilpert, 296–319. Oxford: Routledge.

Perez, C. 1983. "Structural Change and the Assimilation of new Technologies in the Economic and Social System." *Futures* 15: 357–375. doi:10.1016/0016-3287(83)90050-2

Perez, C. 2002. *Technological Revolutions and Financial Capital. The Dynamics of Bubbles and Golden Ages*. Cheltenham: Edward Elgar.

Perez, C. 2009. "The Double Bubble at the Turn of the Century: Technological Roots and Structural Implications." *Cambridge Journal of Economics* 33: 779–805. doi:10.1093/cje/bep028

Reati, A., and J. Toporowski. 2009. "An Economic Policy for the Fifth Long Wave." *PSL Quarterly Review* 62: 147–190.

Scherrer, W. 2016. "Technology and Socio-Economic Development in the Long Run: A "Long Wave"-Perspective." In *Routledge Handbook of Politics and Technology*, edited by U. Hilpert, 50–64. Oxford: Routledge.

Scherrer, W. 2019. "Surfing the Long Wave: Changing Patterns of Innovation in a Long-Term Perspective." In *Diversities of Innovation*, edited by U. Hilpert, 49–68. London: Routledge.

Scherrer, W. 2020. "How "General Purpose Technologies" Trigger Long Waves of Economic Development and Thereby Generate Diversities of Innovation." *UCJC Business and Society Review* 17 (1): 50–65. doi:10.3232/UBR.2020.V17.N1.02

Schuelke-Leech, B. 2018. "A Model for Understanding the Orders of Magnitude of Disruptive Technologies." *Technological Forecasting and Social Change* 129: 261–274. doi:10.1016/j.techfore.2017.09.033

Trotta, D., and P. Garengo. 2018. *Industry 4.0 Key Research Topics: A Bibliometric Review*. 7th International conference on industrial technology and management (ICITM), Oxford, UK.

Xu, L., E. Xu, and L. Li. 2018. "Industry 4.0: State of the art and Future Trends." *International Journal of Production Research* 56 (8): 2941–2962. doi:10.1080/00207543.2018.1444806

Conclusion – the complexity of reorganizing industries and value chains: challenges and selectivity of Industry 4.0 and digitization

Ulrich Hilpert

New technological opportunities associated with Industry 4.0 and digitization have wide influence, in particular, on manufacturing industries and data processing, but also cause change in process industries and many areas of services. While this provides the impression of something fundamentally new in some regions, industries, enterprises, and employees, it is, in fact, a continuation of ongoing research and development in information and communication technologies (ICT), data processing, and transfer, and is about application of such technologies in existing products and improving them for better and more varied purposes. The users of these new and improved equipment and technological opportunities, of course, are confronted with the effects and opportunities at a rather later stage when they are ready to be introduced or shortly before that. These are rarely fundamentally new developments but are substantial improvements based on rather short lead times. Consequently, the cycles of change, improvement, and updates emerge rapidly.

While the technologies are a consequent continuation of ongoing research, there is change in the combination and improvement of such technologies and their application. ICT and data transfer, in particular, allow for networking of equipment, facilities, plants, and jobs. Employees and their workplaces, plants and their demands for materials, services and the data are interlinked. Processes can be organized across locations or even across continents and are integrated into chains. When it comes to modern medical instruments, experts can better participate in or realize remote surgery and make competences available without being physically present at a place. Skills, competences, and collaboration can contribute innovatively to processes and final products. When introducing modern robots they repeat exactly the task expected and make sure that a consistent quality is achieved on time. They may also increase efficiency to an extent that differences in labour costs get marginal, which makes it less profitable to accept long distances of transportation and taking the risks associated with this kind of organization of supply chains.

The higher skills which are required to run such plants may allow relocation to European countries or a highly skilled and educated labour force may provide the basis for such an innovative reorganization. On the other hand, there are opportunities to combine newly developed equipment to reduce the demand for skilled labour while manufacturing quality products. That would make dislocation attractive when combined with investment into new robots. In combination with ICT, the geographical distance does not make it a problem to remotely run and control such plants, neither close to where such supply is required or far away. Regulations of labour or environment, in addition, may make such combinations with suppliers from low-wage countries economically attractive. Thus, it is the new technology and equipment which allows such considerations, despite the various constraints to be considered.

This points to new challenges and opportunities to reorganize manufacturing and related value chains, and it means that the power of these technologies and the equipment associated with Industry 4.0 and digitization provides the basis for organizing a new and changing production system with global impact. How this may perform in future is widely analysed while there are studies about the divergent situations of chains, products, and the capabilities of regions and metropolises. Nevertheless, one may already see the dimensions which matter for participation in these chains, and what it may mean to be placed at a particular point of a chain, when becoming widely dependant on remote management and orders. It is crucial to be prepared for this newly created system and to provide the conditions of participation at the time when this reorganization is achieved.

Due to these various conditions, which form or reorganize a system, there is the perception of a fundamental change. This is, of course, widely correct, but it does not refer to a particular technology. It is important to better understand the systemic relationship between new technological opportunities of developers and those of users, existing industrial capabilities and areas of application, and organizational structures. This requires particular situations, which need to be arranged, and distinguishing between locations, whether they can be linked to such new and dynamic developments and how this can be achieved. In general, the differences between regions and metropolises increase, and whether or not there are linkages with such processes become highly selective. Skills, traditions, particular competences, and infrastructures form the basis of a complexity which requires many elements and contributions at locations and, thus, can be arranged better at metropolises than in less highly urbanized environments. In addition, according to the positioning in global value chains or in international trade, empirically, the complex arrangements vary significantly. Where research and development are based apart there are different skills demanded and specific labour markets are emerging when compared with situations where the various types of suppliers are co-located.

Different from earlier periods of development, in these situations, public policies need to consider more than just certain targets or criteria. It is important to face the specific complexity of a particular region or metropolis in question, and this needs to be understood in relation to the process of development and change. Thus, time is important, because windows of opportunity need to be met while they are open – and which match chains, trade, and location – and each of these elements is subject to constant change, improvement, and new opportunities. Although the far-reaching and overarching impact of Industry 4.0 and digitization demands reorganization at the regional and metropolitan level, the fundamental dimensions of development are still effective and decisive – concerning both opportunities for participation in these processes and improvement of the current situation. Thus, the complexity of both processes and regional or metropolitan situations require an understanding of how to provide for appropriate linkages at a certain moment in time.

While this includes risks when faced with the current complexity, it continues to contain opportunities for innovation and improvement, which are related to tradition, competences, and existing structures. Education, skills, the potential of industries, traditions in particular areas of products, and a relationship to research become even more important than in earlier periods of industrial development. The specificity of such processes is increasing because particular areas of education are required to match with industrial change in a region or metropolis. There is a demand for skills ready to adopt these opportunities while transforming ideas into products and services that are demanded locally and globally when associated with the complexity of chains. Public policies can help build infrastructures for modern high-speed internet and transportation, and they can provide the basis for better skills, education, and research. Whether this may help in innovation and development is related to both the relationships which can be arranged for these new processes and the societal and cultural situations which put enterprises and the workforce into a position to identify its specific opportunities and to engage in the activities which are required during a particular window of opportunity.

Upgrading alongside the value chains and generating higher value-added chains require higher skills and education in manufacturing and services. The application of new technologies or integration into modernized products and processes clearly requires higher skills that suit the patterns of specialization and allow for future development. This helps to establish particular industrial competences and to turn traditional expertise into a clear advantage when it relates Industry 4.0 and digitization to industrial history and previously developed competences. While new areas of research can be established rather easily, this is different from the application of new technologies in relation to complex plants, machines, measurement instruments, or capital goods, and is closely involved with systematic engineering. Consequently, value chains also reflect histories and traditions, which have continued for longer periods of development and several waves of industrial modernization. This also includes the capability to organize manufacturing and supply on time and to provide high quality. As is well known in the aircraft industry in general and the Airbus corporation in particular, reliability and quality are highly important during the organization of a challenging production process. Enterprises and locations, which are known for these competences in production and in products, continue and improve their positions by contributing to the value chain – and thus build innovation-based regional sustainability.

It takes time to build such traditions and they cannot be realized easily. In contrast, infrastructures such as high-speed internet, ICT, and transportation can be introduced quite quickly, as well as new laboratories and the introduction of innovative research by creative academics. Artificial intelligence, software development, or computing are areas which can be established and developed widely with little relation to particular final products. When referring to tradition-based processes versus those based on newly established competences in research, these differences clearly indicate the divergencies among locations which are contributing to value chains. They also point to patterns of specialization and limitations to replicate competences which are based on incremental and location-bound capabilities. The situations arranged at the different locations or metropolises again indicate the complexity which is associated with local situations, as well as when related with value chains.

While the synergy with existing industries, skills, and competences can help to continue industrial development and may arrange for unique selling propositions, there are new locations emerging in relation with research and engineering (e.g. Shenzhen in China, Bangalore in India, or Singapore) which can provide future areas of specialization and lay the ground for emerging particular traditions. Such processes of development demand a critical mass of knowledge workers and established situations in research and development. This takes quite some time, but much can be arranged within a foreseeable period of time. It is associated with socio-economic change and patterns of organization which suit the situations of regions and metropolises and needs to be combined with modern infrastructures. Consequently, geography matters when many industrial and intellectual capabilities are enabled with such facilities, and thus become linked to global chains and progress. Metropolises provide such situations and are addressed with priority because the modernization of infrastructure will enable many potentialities to contribute to value chains, to develop and apply new technologies, to supply advanced services, or to manage plants remotely.

This attracts people for work and has an effect on the concentration of capabilities, as well as changing of metropolitan societies with an enhanced concentration of knowledge workers. Accordingly, when applying Industry 4.0 and digitization the organizational patterns for particular value chains change and take into consideration the diversities of the metropolises embedded in their networks. The combination of new or advanced technologies with modern infrastructures also allows improved efficiency at lower levels of the value chains by both becoming more closely linked with global chains and by creating the opportunity to introduce efficient robots in cooperation with low-skilled and cheap labour. The main competence in engineering, technology, and management, of course, can be contributed remotely by new ICT, which either continues to rely on low levels of knowledge workers or may even reduce this level by replacing them remotely from abroad. This virtual presence of experts and their exchange via new media may even make these competences more effective as they can be related more easily to a larger number of locations or metropolises and by a concentration at a few places which provides for easy exchange of knowledge and experience among managers and high-level engineers.

While this tends to concentrate competences and to make management and technological assistance virtually available, it also means that certain competences and decision-making by management are reduced at the local plants and in the region. Smart factories are examples of reduced management competences at these locations, which also means that there are less frequent opportunities for collaboration among the potentialities and capabilities of different areas. Since the chains are increasingly organized and run centrally, the specific advantages of locations are addressed and supported by centralized competences and, thus, regionally there is only a limited complexity which needs to be managed. There is a kind of synergy based on the organization of value chains, and a close relationship is introduced between particular knowledge contributions and the regions where these tasks are realized. This division of labour alongside the chain and a tailorization of the processes helps to organize the contributions which are generated with limited complexity at the individual location because it focuses only on elements of the whole system. The concertation of these contributions is realized by organizing the value chain and the integration of semi-finished goods into the final product or the final results of research and services provided.

Consequently, important parts of the processes are realized outside of a region or metropolis, and decisions which are highly important for these locations are taken elsewhere. Consequently Industry 4.0 and digitization highlight the particular importance of the complexity of organization. This indicates the need to take into consideration the entire process when preparing for regional or metropolitan development. The rationality of multinational enterprises and their interest in particular locations widely influence locations, suppliers, and the opportunities of clusters. Vice versa, strengthening regional competences can make it attractive for large enterprises to take advantage of such locations, can link research to a global scientific community to share new findings in research networks, or to become an attractive location for start-ups. Thus, this kind of innovation also allows for new opportunities when performing attractively to many different users and value chains. The rationality of global value chains also provides for access to global markets and building a necessary competence in research, manufacturing, or industry-related services can link such clusters to dynamic processes.

Since there is a rich number of value chains which can take advantage of such competences, the dependency on a chain can also turn into an attractive situation regarding different value chains. Such converging recognition of an attraction of a location is already well known when considering labour-intensive products exploiting cheap labour in various areas of manufacturing. But such locations, which are strong in innovative and highly reliable suppliers, makes them strategically important and difficult to be replaced. Industry 4.0 and digitization can strengthen their position and can add a new perspective to the complexity of value chains when placing them into a position of increasing relevance for the system itself and its organization. Such clusters or individual suppliers are not characterized by chain dependence, rather the chains are forming their markets. The diversities of products, chains, and processes in a global situation means that the context of individual regions, metropolises, industries, and enterprises need to be considered when these new technological opportunities are analysed. There is a clear relationship with the context which needs to be considered. Innovation and industrial opportunities, regional and metropolitan development, competences and research capabilities are highly divergent and need to be understood in the light of the relationship which exists between the situation they form and their wider context.

This raises new questions concerning the suitability of divergent processes of innovation and what opportunities exist in different regions or metropolises. Although this seems to be rather clear for different industries and chains concerning individual locations and their contextualization, the systemic relationship still needs to be demonstrated. Similarly, when discussing their divergent opportunities for regional development the contextualization of high-competence clusters needs to be demonstrated. Finally, Industry 4.0 and digitization can help for better understanding the relationship between industrial innovation systems and regional innovation systems, because the activities of industries, enterprises, and chains are geographically located.

Index

Note: Page numbers in **bold** refer to tables, page numbers in *italics* refer to figures and those followed by 'n' indicate notes.

aerospace and automotive industries 26–39; adding new regions into supply chain 32–3; character and impact of Industry 4.0 26–30; innovation in supply chains 36–7; new entrants 33–6; opportunities 33–6; reinforcement of established regions 30–2

Albizu, E. 47

artificial intelligence (AI) 1, 27, 44, 160; for activity automation 84; autonomous robots and 80; and digital technology 97, 151–2, 155; in metropolitan areas 107n1; and remote diagnosis/maintenance 85; supportive technologies 98; trained software engineers 36

augmented reality (AR) 148; in aero-space MRO 28; for manufacturing workers 27; optimization of processes 11

automation 94–8; artificial intelligence (AI) for activity 84; Europe 94–106; regionalized variations 95–6; risks 99–106; United States 94–106

automotive industries *see* aerospace and automotive industries

autonomous robots 80, 148

Baden-Württemberg, Germany 2, 5, 21, 60, 77–8, 81–8, 94

Barker, R. 94

Basque Country and Catalonia 42–54; bipolar innovation policies *vs.* skill deficits 48–50; convergencies in skill deficits 50–3; divergent innovative opportunities 50–3; Industry 4.0 in Spain 43–4; policy instruments to promote Industry 4.0 in 47, 50; regional innovation 53; role of existing structures in industry and institutions 44–7

Basque Digital Innovation Hub 44–5, 47

Basque Fifth Vocational Training Plan 46

Basque Industry 4.0 44–5, 47, 54n3

Berube, A. 94

big data: and advances in machine learning 27; analytics 148; manipulation 11

Bind 4.0 (public-private startup accelerator) 46, 47

bipolar innovation policies 48–50

Brandenburg, Germany 2, 21, 35, 77, 81–8

Buhr, D. 107n4

Calenda, Carlo 63

Carvalho, N. 113

Catalonia *see* Basque Country and Catalonia

Ceruti, A. 28

challenges: economic 76–7; to existing processes and value chains 11; of Industry 4.0 technologies 28–30, 50; and selectivity 158–62

change 143–55; divergent regional perceptions of disruption 151–2; not-so-sudden appearance of (much of) Industry 4.0 147–52, 154–5; relevance and disruptiveness of 147–9; 'stand-alone' perspective 144–7; 'sudden' 144–7; unexpectedness and perception of change 149–51

cloud services 148

collaborative synergies 128–30, 137

competences 20–1; capabilities and 19; divergent 3, 7, 9, 18; human 2; industrial 4–5, 10; regionalization by 12–15; research capabilities 162

computer integrated manufacturing (CIM) 115

context matters 137

COVID-19 pandemic 139

cyber physical systems 147

cyber-physical systems (CPS) 43, 89n1

cyber security 148

decision-making 42, 48, 116–17

Digital Index 80

disruption/disruptive: divergent regional perceptions of 151–2; innovation 89n3; relevance and 147–9; technology 101

disruptiveness of change 145, 147–9

divergent innovative opportunities 50–3

3D-printing 84–5, 148

Drath, R. 148

Electronics and Telecommunication Research Institute (ETRI) 125n1

employees, consequences for 67–8

engineering industry 70n1
entrants, new 33–6
Europe: automation 94–106; effects of
    digitalization and regional variations 94–5;
    metropolitan medium skill 97–8; national
    economic development 98–9; regionalized
    automation variations 95–6; skills misfit 94–8;
    systematic sectoral variations 98–9; worker
    displacement 94–8

First Industrial Revolution 155n1
flexible manufacturing systems (FMS) 114–15
Fraunhofer Industrie 4.0 Layer Model 113–
    14, 113
Freeman, C. 112

Gaddi, M. 148
Ganzarain, J. 148
Garbellini, N. 148
Garengo, P. 147
Garibaldo, F. 148
General Data Protection Regulation 150
general purpose technologies (GPT) 147–8, 151–2
Germany 76–88; Baden-Württemberg 81–8;
    Brandenburg 81–8; divergent opportunities 79–
    81; federal political strategy 78–9; Industry
    4.0 78–9, 81–8; manufacturing sector 101;
    North-Rhine-Westphalia 81–8; problems in
    manufacturing industries 79–81; regional
    disparities 79–88; states of 89n2; varying
    significance of the industrial sector for 76–8
government plans/program/policies: in case of
    regional capabilities 18–22; Italy 63–6; regional 46

Hanover Fair 2011 79
Horch, A. 148
horizontal integration 148

IAB (Institut für Arbeitsmarkt-und
    Berufsfroschung) 89n11
Ibarra, D. 148
Igarta, J. 148
IG Metall (Industriegewerkschaft Metall) 89n9
    n. 9–10
industrial clusters 137
industrial Internet 147
Industrial Internet of Things (IIoT) 129
4th Industrial Revolution (Korea) 112–19, 125n1,
    125n3; see also Korea
industrial revolutions 144–7
industrial triangle 60
Industry 4.0, origin of the term 88–9n1
industry-location-nexus 12–15
inequality: income 98; reducing 153; regional in
    Italy 66–7
information and communication technologies
    (ICTs) 110; competitiveness 78; industry 133–4;
    infrastructure 139; paradigm 149; revolution 27;
    technology 64, 113

innovative processes 9–23; facing divergent
    situations 15–18; government policies in
    case of regional capabilities 18–22; industry-
    location-nexus 12–15; regionalization by
    competences 12–15; regionally divergent
    adaptation to new opportunities 10–12;
    societal tendencies of technological
    opportunities 22–3
Internet of Things (IoT) 11, 147
Italy 57–70; consequences for employees 67–8;
    government plans/program 63–6; industrial
    structure 57–60; Italian Industry 4.0 62;
    manufacturing industry 62–3; patterns of
    change 67–8; regional inequality 66–7;
    relevance of the societal structures 60–2; social
    aspects 68–9

Kondratieff, N. 111–12
Korea 110–25; 4th Industrial Revolution
    112–19, 125n1, 125n3; Industry 4.0 112–15,
    114, 124–5; Korea Smart Manufacturing
    Office (KOSMO) 120–1; potential to respond
    from regional perspectives 117–19; smart
    factories 119–24, **121–2**, 123; uneven regional
    development 124–5

Li, L. 148, 155n1
Liao, Y. 147

M2M (Machine to Machine) 43
manufacturing industry: Italy 62–3; problems in
    Germany 79–81; revenue in Germany 80
manufacturing systems 147
Massachusetts 134–5
MassTech Intern Partnership 135
Maxim, R. 94, 99–100, 108n6
McKinsey Global Institute 94
mechatronics 61
MES (Manufacturing Execution Systems) 67
metropolitan medium skill 97–8
Metropolitan Statistical Areas (MSAs) 99
Ministry of Science and ICT (MSIT) 116, 120
Ministry of SMEs and Startups (MSS) 116, 120
Ministry of Trade, Industry and Energy (MOTIE)
    116, 120
Mobility as a Service 29
MRP (Material Requirement Planning) 67
Muro, M. 94, 99–100, 108n6

national economic development 98–9
National Industrial Strategy 2030 80
national innovation system (NIS) 115
national policies 152–4
network intermediaries 137–8
Neugebauer, R. 113–14
North, K. 47
North America 128–39; approach
    (method) and considerations 130–2;
    collaborative synergies 137; context

matters 137; industrial clusters 137; industry 4.0/digitalization and innovation networks 130; Massachusetts 134–5; network intermediaries 137–8; Ontario 135–6; regional structures 130; results/analysis 132–6
North-Rhine-Westphalia, Germany 77, 79, 81–8

Ontario 135–6
opportunities: aerospace and automotive industries 33–6; divergent innovative 50–3; Germany 79–81; regionally divergent adaptation to 10–12; societal tendencies of technological 22–3
original equipment manufacturers (OEMs) 11, 14, 16, 19–21, 26, 38–9, 134; aerospace 34; automotive 33; and design and build suppliers 31; and digital suppliers 36; EU value chain with 62; innovative impulse of global 43; in well-established regions 32

perception of change 146, 149–51; *see also* change
Perez, C. 112, 144, 147–8
Platform Industry 4.0 83
Presidential Committee on the 4th Industrial Revolution (PCFIR) 111–12, 116–17, *117*, 124
production processes 147
production systems 147
Program for Diffusing and Advancing Smart Factory 119
PwC 28, 31, 33

regional innovation **53**
regionalization by competences 12–15
regionalized automation variations 95–6
regionally divergent adaptation 10–12
regional variations 94–5
Reischauer, G. 51
reorganizing industries 158–62
research and development (R&D) 28, 67, 76, 83, 86, 88, 97, 130–1, 138, 154, 158–9
Rifkin, J. 112
RIS3CAT 2015–2020 Action Plan 49

Schumpeter, J. 112
Schwab, K. 111–12
Second Industrial Revolution 155n1
selectivity and challenges 158–62; *see also* challenges

simulation 148
skill deficits 48–50
skills misfit 94–8
Small and Medium Business Administration (SMBA) 120
small and medium-sized enterprises (SMEs) 33, 36, 49, 80, 134; Basque 47; competitiveness of 125; in Italy 70; in Korea 119–24; local industries and 48; in national and regional strategies 62; Valencian 44
smart factories 27, 147; Korea 119–24, **121–2**, *123*
smart manufacturing 147
Spain 54n1 n. 1–2; Industry 4.0 in 43–4
'sudden' change 144–7; *see also* change
supply chains: adding new regions into 32–3; innovation in 36–7
systematic sectoral variations 98–9

tax credit 66–7
techno-economic paradigm (TEP) 147–55
technological change 29, 33, 51, 144, 152
technological innovation 15, 28, 67, 69, 112, 130
telemigration 27
Theorin, A. 28
Third Industrial Revolution 155n1
Tomer, A. 107n1
Trotta, D. 147

unexpectedness: of change 145–6; and perception of change 149–51
United States 92–107; automation 94–106; effects of digitalization and regional variations 94–5; metropolitan medium skill 97–8; national economic development 98–9; regionalized automation variations 95–6; skills misfit 94–8; systematic sectoral variations 98–9; worker displacement 94–8

value chains 158–62
vertical integration 148

Whiton, J 99–100, 108n6
worker displacement 94–8
World Economic Forum 27

Xu, E. 148, 155n1
Xu, L. 148, 155n1